Ecosickness in Contemporary U.S. Fiction

LIVERPOOL JMU LIBRARY

LITERATURE NOW

Matthew Hart, David James, and Rebecca L. Walkowitz, *series editors*

Literature Now offers a distinct vision of late-twentieth- and early-twenty-first-century literary culture. Addressing contemporary literature and the ways we understand its meaning, the series includes books that are comparative and transnational in scope as well as those that focus on national and regional literary cultures.

Toward the Geopolitical Novel: U.S. Fiction in the Twenty-First Century,
CAREN IRR

ECOSICKNESS
in Contemporary U.S. Fiction

Environment and Affect

HEATHER HOUSER

Columbia University Press New York

Columbia University Press
Publishers Since 1893
New York   Chichester, West Sussex
cup.columbia.edu
Copyright © 2014 Columbia University Press
Paperback edition, 2016
All rights reserved

Library of Congress Cataloging-in-Publication Data
Houser, Heather.
Ecosickness in Contemporary U.S. Fiction. Environment and Affect /
Heather Houser
pages cm. — (Literature Now)
Includes bibliographical references and index.
ISBN 978-0-231-16514-3 (cloth : alk. paper)—ISBN 978-0-231-16515-0
(pbk. : alk. paper)—ISBN 978-0-231-53736-0 (e-book)
1. American literature—History and criticism. 2. Environmentalism in
literature. 3. Diseases in literature. 4. Ecocriticism. I. Title.

PS169.E25H68   2014
810.9'36—dc23   2013041366

Columbia University Press books are printed on permanent and durable
acid-free paper.
Printed in the United States of America

Cover design by Julia Kushnirsky
Cover illustration by Olaf Hajek

TO BETTE
AND
JAY (1948–2009)

CONTENTS

Acknowledgments ix

1. Ecosickness 1
   Sickness in a Technoscientific Age 8
   Life, Ethics, and Action 12
   Ecosickness in the Field 19
   Outline of the Book 27

2. AIDS Memoirs Out of the City: Discordant Natures 31
   Prologue 31
   Contested Natures 39
   North Enough's "Difficult Beauties" 46
   The "Con" in *Close to the Knives* 55
   Discordant Feelings, Suspicious Stances 65
   Discord in Activism 72

3. Richard Powers's Strange Wonder 77
   "Weirdly Alive" with Wonder 81
   "The Ordinary by Another Name" 93
   "Struggling with Complex Interactions" 100
   "The Ethic of Tending" 109

viii CONTENTS

4. *Infinite Jest*'s Environmental Case for Disgust   117
    Detached Dispositions   124
    "Experial" Ambitions   130
    Body Building   139
    Affective Itineraries   145
    How to Do Things with Disgust   152

5. The Anxiety of Intervention in Leslie Marmon Silko and Marge Piercy   167
    Disrupting the "Pattern of Disease"   170
    "A Single Configuration" of Land and Body   175
    Iniquitous Interventions   185
    Anxious Apocalypse   195
    Squirming and Trembling   208

Conclusion: How Does It Feel?   217

Notes   229
Works Cited   269
Index   295

ACKNOWLEDGMENTS

Writing a book is an exercise in scholarly collaboration, even when only one author's name appears on the cover. This project came to be under the guidance of stalwart teachers, most notably Ursula Heise, Gavin Jones, David Palumbo-Liu, and Nicholas Jenkins. I have learned and benefited from their deep knowledge, keen eye for argument, commitment to making the bold claim, and attention to the nuances of language and narrative. Ramón Saldívar and Sianne Ngai also enriched this project and offered invaluable counsel. For setting me on the path that led me here, I'm grateful to Bill Ray.

At the University of Texas at Austin my colleagues—and friends—have helped me and this project grow with their ready encouragement. Chad Bennett, Gabrielle Calvocoressi, Evan Carton, Neville Hoad, Susannah Hollister, Cole Hutchison, Julia Lee, Lindsay Reckson, Matt Richardson, and Snehal Shingavi provided fresh perspectives on chapters of this book. Cole and Matt Cohen helped me navigate the publishing process. I am grateful to my chairperson, Liz Cullingford, and associate chairperson, Martin Kevorkian, for creating the productive and affable work environment in which a young scholar can thrive. My mentors, Jennifer Wilks and Evan, have also made the first years on the UT faculty fruitful ones. I valued the opportunity to share this work at the Center for Women's and Gender Studies and thank Sue Heinzelman and fellow members of the 2011-2012 Faculty Development Program for their audience. Andrea Golden and Cecilia Smith-Morris help keep the ship afloat. Without Colleen Eils's hard work, deadlines would have melted away along with my sanity.

Allison Carruth encouraged me early on to let my scholarship speak my passions. Thank you. Dear friends Claire Bowen, Joel Burges, Justin Eichenlaub, Harris Feinsod, Michael Hoyer, Ruth Kaplan, and Ju Yon Kim have seen this project in its many stages and have been my brilliant interlocutors (often over whiskey or wine) for years. Kiara Vigil continues to be a close but "outside" reader. Conversations at Rice University, UC Davis, Macalester College, Williams College, and Stanford's Center for the Study of the Novel and Contemporary Reading Group, as well as conferences for the Association for the Study of Literature and Environment and the American Comparative Literature Association stimulated thought on this book. Priscilla Wald, Sam Cohen, and Lee Konstantinou helped the ideas found herein get an early audience.

Rebecca Walkowitz expressed curiosity about the mind behind this project and brought the book to Matthew Hart and David James, coeditors of the Literature Now series at Columbia University Press. She is a great advocate. At the Press, it has been a privilege to work with editor Philip Leventhal. I also thank the anonymous readers for their critiques, Susan Pensak and Audrey Smith for attention to the manuscript and production, and Whitney Johnson for her patience.

I am grateful for the financial support I have received at all stages of this project, including the following from the University of Texas at Austin: a Book Subvention Grant from the Office of the President, a Summer Research Assignment from the Graduate School Faculty Development Program, and a Center for Women's and Gender Studies Faculty Development Program Fellowship. Support also came from a Mellon Postdoctoral Fellowship at the Center for Environmental Studies at Williams College, the Mrs. Giles Whiting Foundation, Stanford University, and the U.S. Department of Education Jacob K. Javits Program.

An earlier version of chapter 3 appeared as "Wondrous Strange: Eco-Sickness, Emotion, and *The Echo Maker*," *American Literature* 84, no. 2 (2012): 381–408, copyright 2012, Duke University Press. All rights reserved. Reprinted by permission of the publisher, Duke

University Press (www.dukeupress.edu). An earlier version of chapter 4 appeared as "*Infinite Jest*'s Environmental Case for Disgust" in *The Legacy of David Foster Wallace*, ed. Samuel Cohen and Lee Konstantinou (Iowa City: University of Iowa Press, 2012), 118–142. Reprinted by permission of University of Iowa Press.

Above all, and every day, I thank my mother, Bette Houser. Her love, indomitable will, and appreciation for joy have always guided me. I could never be grateful enough, but know that it's always enough. Garrett Houser: we bitch and moan, reminisce, and even brag a little, but you continue to keep me in my place. My love to you both.

Ecosickness in Contemporary U.S. Fiction

1

Ecosickness

Carol White and Laura Bodey are under assault. These white, middle-class women inhabit contrasting geographies of the late twentieth-century United States—a manicured suburb of Los Angeles and a nowhere town in middle America—but these distinctions melt away when chemicals infuse them. Cumulative exposures to everyday products like perfumes, dry cleaning fluid, shampoos, and new furniture that were meant to beautify and sanitize their lives have instead poisoned them. Carol and Laura diminish moment by moment, and they are cut off from the future. Chronic rather than terminal, Carol's environmentally induced ailments are incrementally debilitating rather than deadly. Asthma attacks, headaches, stinging red eyes, hive outbreaks, and seizures all add up to a probable but uncertain diagnosis of multiple chemical sensitivity (MCS), the paradigmatic syndrome of "our chemically troubled times."[1] Laura's diagnosis of ovarian cancer is more perilous, but it too correlates to constant chemical exposure from household consumer goods. Even though these two women belong to national and socioeconomic populations that are least vulnerable to environmental illness, according to the World Health Organization, Carol and Laura carry the "disease burden attributed to key environmental risks globally and regionally" in the 1990s.[2] The long-term damages of industrialization, specifically chemical production, are manifesting violently not only in expected victims such as workers who handle hazardous materials but also in mothers whose domains are the kitchen, the garden, and the supermarket.

Told in Todd Haynes's film *Safe* (1995) and Richard Powers's novel *Gain* (1998), respectively, Carol's and Laura's stories carry powerful emotions as they alert audiences to conditions of toxic endangerment that impact contemporary Americans. The monosyllabic simplicity of the titles *Safe* and *Gain* prepares us for straightforward plots of exposure and resulting illness, but the actual narratives are anything but tidy. They implicate mundane commodities such as Pam cooking spray and Dawn dish soap in somatic illness and elicit horror, in the case of Haynes's film, and sentimentality, in Powers's novel. Yet both texts refuse neat causal explanations for these women's medical decline, and their affective power hinges on the uncertainties they sustain. Carol "can't help it" that she can no longer bear life in the San Fernando Valley, and medical science can't help her much either.[3] Residues of doubt always remain as Carol attempts to pin down the origins of her draining symptoms, just as residues of hair spray linger in the locker room at her gym. Laura's diagnosis, by contrast, is indisputable: tests show that she has cancer, and medical specialists can track her diminishing immune cell count. Nonetheless, *Gain* does not close Laura's case; it never identifies the protagonist's chemical ambience as the empirical source of her cancer.

Though the plots of *Safe* and *Gain* withhold causality, they revolve around tenacious searches for the lines that will connect environmental toxification to human illness. This book emerges from the interest in pervasive sickness that inspires Haynes, Powers, and environmental thinkers more generally, but it asks what happens when artists abandon quests for etiology as the driving force of their narratives. I argue that an emergent literary mode, "ecosickness fiction," comes to the fore to join experiences of ecological and somatic damage through narrative affect. The most basic point that *Ecosickness in Contemporary U.S. Fiction* makes is that contemporary novels and memoirs deploy affect in narratives of sick bodies to bring readers to environmental consciousness. The texts gathered here eschew causal models for representing human bodies enmeshed in their environments and instead posit the interdependence of earth and soma through affect. As I will go on to describe in more detail, the meanings of affect are legion within contemporary cultural studies and the disciplines of philosophy and psychology on which it often draws. My assumption in this book is that "affect" designates body-based feelings that arise in response to elicitors as varied as interpersonal and institutional relations, aesthetic experience, ideas, sensations, and material conditions in one's environment. Though there is a relation between affect and eliciting conditions, the relation is not determinate. That is, the same elicitor can excite different affects in different people, and sometimes affect has no specifiable catalyst. The embodied and the cognitive mingle in affect. The feeling grounds one in the present, but it is also coded by past experience and impinges on the future. Indeed, while I agree with Sianne Ngai that some affects are "less than ideally suited for setting and realizing clearly defined goals," affect in general positions us to adjust modes of thought, to act (or remain passive), and to make decisions.[4] In short, it is at the root of our social, political, and ethical being and thus, this book argues, at the root of environmental orientations. The phrase *narrative affect* abbreviates my argument that affects are attached to formal dimensions of texts such as metaphor, plot structure, and character relations.

The chapters of this book theorize the formal strategies that become engines of affect in contemporary fiction and that authors use to imaginatively understand sick life. In ecosickness fiction, humans and the more-than-human world do not only interact but, more importantly, are coconstitutive. This literature shows the conceptual and material dissolutions of the body-environment boundary through sickness and thus alters environmental perception and politics. Uniting earth and soma through the sickness trope, albeit a trope with a material reality, ecosickness narratives involve readers ethically in our collective bodily and environmental futures. Serving these functions, sickness organizes many of the thematic, ethical, formal, and affective investments of late twentieth- and early twenty-first-century fiction. With sickness as an analytic framework, the readings to follow show that recent novels and memoirs develop narrative affects that draw conceptual homologies between environmental and somatic vulnerability. It is through affect that recent fiction envisions the shared endangerment and technologization of contemporary bodies and environmental systems. For this reason, powerful but unpredictable affects drive the chapters of this book just as these affects drive ethical responses to forms of endangerment—from habitat decimation and species loss to depression and substance addiction, from pollution and suburban sprawl to HIV/AIDS and MCS.

How do contemporary writers imagine embodied engagement with environments and reconceive ethical relations with the more-than-human? A cross-section of recent literary production helps us answer this question and includes authors Jan Zita Grover, David Wojnarowicz, Powers, David Foster Wallace, Marge Piercy, and Leslie Marmon Silko. These writers share two main approaches to relating text and world as they create a sickness imaginary. First, they attend to how contemporaneous scientific researchers, medical professionals, activists, and policy makers are reimagining "life itself" and then contribute their own conception of biological life as malleable and vulnerable. The seemingly mundane phrase *life itself* has a history. As Eugene Thacker reminds us, it started to dominate popular writing on molecular biology in the 1950s and 1960s. It disseminated the idea "that there was a master code that coded for the very biological foundations of life," even as the search for the essence of life remained—and remains—elusive.[5] "Code" converts genetics into a branch of informatics and makes biological life available for instrumental, profit-driven manipulation. Second, even as these writers learn from quantitative and technical disciplines, they do not put all of their eggs in the basket of empiricism and technological rationality. Instead, ecosickness fiction attests that an array of stories and narrative affects is necessary for apprehending the material and conceptual relays between the embodied individual and large-scale environmental forces. I take methodological inspiration from ecosickness writers who draw on fields of knowledge as diverse as neurophysiology and the history of the Americas. I examine the fictional and nonfictional writings of these authors, as well as some orthogonal discourses that nourish them: advertising, science writing, visual art, popular journalism, and activist campaigns.

Taken together, ecosickness narratives establish that environmental and biomedical dilemmas produce representational dilemmas, problems of literary form that the techniques of postmodernism, realism, nature writing, scientific communication, or activist polemic alone cannot neatly resolve. Though there is no one dominant tradition guiding ecosickness fiction, it undoubtedly has a prehistory and a shadow, that is a corpus of contemporary texts that double its interests but do not entirely overlap with them. Rachel Carson's *Silent Spring* (1962) deserves pride of place in the prehistory of ecosickness fiction. Notable for its influence on postwar environmental policy and activism and for the literariness of its introductory "A Fable for Tomorrow," *Silent Spring* sought to convince readers that animal and human bodies are barometers of ecosystemic toxicity, a lesson that my authors have internalized and work to impart to their audiences. This prehistory also includes Meridel Le Sueur's stories from the 1940s that, in Stacy Alaimo's analysis, offer a "spiraling imbrication" of material bodies and earth that is routed through labor and class inequalities.[6] Stories such as "Eroded Woman" (1948) announce the "many strains of melancholy" that sing out from an Oklahoma town wasted by lead mining.[7] "'Nothing will ever grow,'" laments the titular woman as the effluvia of mining replace local flora, and it "'seems like you're getting sludge in yore [sic] blood.'"[8] Upton Sinclair's kindred project in *The Jungle* (1906) inspires ecosickness authors as it incorporates the city into the U.S. literary environmental imagination and puts bodies under the influence of the air they breathe and the fellow creatures they kill. Just as Le Sueur's eroded woman can't see the forest for the waste of mining, "one never saw the fields, nor any green thing whatever" in the shantytown of Sinclair's Chicago.[9] Le Sueur and Sinclair confront the question "was it not unhealthful" to breathe in and soak up the detritus of industrialized production; they thereby introduce the powerful relays between soma and space that their successors depict under more advanced conditions of pollution, urbanization, and technologization.[10]

As I will describe more fully, these advanced conditions distinguish post-1970s ecosickness fiction from its antecedents. In the texts that I examine here, changes to the matter of life—"sludge in yore blood," as the eroded woman explains it—are not only the unexpected by-products of industrial mining and agricultural practices. They also arise from technoscientific ventures that intervene in life itself and change the very matter of being. Heightened technologization and medicalization of body and earth have placed two questions before contemporary authors: how do interventions into the very stuff of life make us *feel*? And how do those feelings reconfigure environmental and biomedical ethics and politics? What interests me are authors who approach these questions without using causality as a motivating logic.

This point brings me to those works that shadow the concerns of ecosickness fiction but do not fall under the rubric. First, readers might expect to find Don DeLillo's *White Noise* (1984) in the forthcoming chapters. *White Noise* narrates the epistemological uncertainties (and absurdities) of contemporary sickness through responses to an "airborne toxic event" and to a lifestyle drug called Dylar, which is designed to mute the dread of death and life in a society of risk.[11] However, DeLillo's novel is not central to my arguments here because somatic sickness is primarily anticipated and simulated rather than lived by its characters. Therefore, embodied experience of sickness does not provoke its environmental arguments and biomedical imagination as it does for the authors of ecosickness. While *Safe* and *Gain* do materialize sickness, they, along with a plethora of memoirs of environmental toxicity,[12] are not centerpieces of this project because etiology is a specter that haunts their plots. Questions of origins preoccupy their narratives even as—or because—origins are ambiguous. I direct critical attention to narratives in which environmental and somatic sickness correlate conceptually, affectively, and imaginatively, but where environmental factors and disease are not perforce etiologically related.

Why think outside of causality? After all, shouldn't artists and critics join activists, journalists, and scientists in "connecting the dots" between environmental contamination and syndromes like MCS, infertility, and mental illness? I examine recent cultural production as making visible these connections, but in a manner that is independent of but in dialogue with scientific communiqués and unique to artistic works. Literature's contribution to this dialogue comes through most compellingly when it brackets causes and the empirical approaches that isolate them. This does not mean that ecosickness fiction is antiscience. While authors such as Silko direct strong skepticism toward the sciences of life and others such as Powers critique the instrumentalism and reductionism of certain clinical methods, the texts that comprise this project have a broader aim: to approach scientific research as an avowedly shifting foundation for knowledge and to promote alternative epistemologies of emotion and of narration. In this respect, ecosickness fiction agrees with Wai Chee Dimock and Priscilla Wald who advise that "the practical impact of . . . specialized knowledge—from reproductive technologies to electronic archives, from bioterrorism to gene therapy—makes science illiteracy no longer an option" for humanists.[13] But the stronger statement that the authors assembled here make is that *narrative* illiteracy is no longer an option for the environmental and biomedical citizens we are called to be. Apprehending planetary and physiological sickness requires literary and more broadly humanistic knowledge. To this end, this literature brings body and earth together through narrative affect to illuminate how emotion rather than empiricism alone powerfully, if not always predictably, conducts individuals from information to awareness and ethics. With attention to narrative affect, we at once establish that the embodied person is enmeshed in macro processes of technologization and environmental manipulation and acknowledge that awareness of enmeshment does not dictate a singular ethics or politics. Rather, this awareness invites further cultural experimentation with how emotions are resources for ethical stances and political action.

Each chapter of this book analyzes how a particular affect contours a narrative of environmental investment or disengagement that is centered on sick bodies. Discord, wonder, disgust, and anxiety are the affects that animate works by Grover and Wojnarowicz, Powers, Wallace, and Piercy and Silko, respectively. These emotions have a central place in ecosickness fiction either because they have traditionally shaped environmental and medical enthusiasm, as wonder and anxiety have, or because they seem to oppose the modes of attachment that environmental thought articulates, as with discord and disgust. The texts I gather here provide strategies for coping with the environmental and bodily threats that preoccupy artists as well as scientists, environmentalists, and policy experts. Specifically, they mobilize affects that variously aid or inhibit ethical involvement in entwined dimensions of contemporary U.S. existence: the reimagining of life itself, the medicalization of life, and ecological endangerment.

Ecosickness argues that responses to these shifts emerge through the stories we tell, the metaphors we employ, the forms of relatedness we envision, and the emotions they all produce. In particular, it urges environmental humanists and ecocritics in particular to attend to the intimacy not only of planetary and bodily injury but also of narration, affect, and ethics. Affect is pivotal to the complexity of emergent concerns about climate change, species extinction, pervasive toxicity, population growth, capitalist expansion, and technoscientific innovation. As Paul and Scott Slovic have announced, "we need numbers and we need nerves" because, "without affect, information lacks meaning and will not be used in judgment and decision making."[14] This literature does not, however, articulate univocal statements of judgment and decision making. Rather it entertains the affective appeals of a variety of ethical stances—laissez-faire, anti-interventionist, posthumanist, preservationist, and even nihilist—and how these stances then dispose us toward political action or "mere" survival.

SICKNESS IN A TECHNOSCIENTIFIC AGE

The historical footprint of *Ecosickness in Contemporary U.S. Fiction* is both small and near, with most texts published in the 1990s and early 2000s. However, the nexus of concerns that inspire the *fin de millénaire* literature of sickness reaches back to the 1960s and 1970s, when my earliest text was published. *Silent Spring* ushered in restrictions on pesticides and heralded an era of heightened regulations such as the Clean Air (1963), Wilderness (1964), and Clean Water (1977) acts. This legislation announced the precariousness of ecosystems at the same time as it assigned to industry, municipalities, and government agencies some responsibility for preserving biotic systems.[15] With the first Earth Day in 1970 and NASA's 1972 broadcast of an image of the full earth as seen from space (known variously as the Blue Planet, Blue Marble, and Spaceship Earth), these decades also inaugurated a new planetary consciousness.[16] In the 1990s U.S. environmentalism mutated as efforts to prevent ozone depletion and research into global warming began shaping the movement. Environmentalists began addressing how technological growth intensifies environmental degradation and might also mitigate it. A central technology in this debate has been genetic modification, which entered environmentalism through considerations of whether feeding the world's growing population requires widespread use of genetically altered seeds, plants, and other organisms.

The rise of genetic technologies draws together environmentalist and biomedical discourses at the end of the twentieth century. Along with an evolving sense of planetary responsibility, ecosickness fiction takes stock of how the popularization of genetic science transforms people's conceptions of life itself. The invention of recombinant DNA techniques (or gene splicing) in 1973 and attendant forms of creativity within the molecular biological and biomedical sciences opened up avenues to reconfiguring life matter.[17] For sociologist Nikolas Rose, what makes technologies such as regenerative medicine, gene therapy, and genetic modification novel is that they "do not just seek to cure organic damage or disease, nor to enhance health, as in dietary and fitness regimens, but change what it is to be a biological organism, by making it possible to refigure—or hope to refigure—vital processes themselves."[18] They are thus integral to the shift from a medicalized society to a biomedicalized one. Under medicalization, which occurs between the 1940s and 1990s, "aspects of life previously outside the jurisdiction of medicine come to be construed as medical problems."[19] "Particular social problems deemed morally problematic and often affecting the body (e.g., alcoholism, homosexuality, abortion, and drug abuse) were moved from the professional jurisdiction of the law to that of medicine," and the "medical industrial complex" penetrated the economic and social spheres more completely.[20] Biotechnologies, computerization, and the privatization of medical research and care enhance the shift to biomedicalization at the tail end of the twentieth century. For the purposes of my argument, the biotechnological dimension of this trend is of utmost importance as innovations such as gene therapy do not only stave off disease but so profoundly alter the parameters and makeup of life that they change conceptions of it.[21] Under biomedicalization the "harnessing and transformation of internal nature (i.e., biological processes of human and nonhuman life forms)" eclipse control of outside forces as the goals of medical research and services.[22] "Optimization," or enhancement of the body, attracts medical expertise, and the patient is frequently rebranded as a client or consumer. Optimization has two main consequences. First, even as the human body is seen to be perfectible, it never measures up to perfection. Asymptotically approaching a norm of optimal performance and beauty, the human body thus becomes increasingly available to manipulation and intervention. Second, individuals in the West come to understand themselves as biomedical subjects, responsible for managing personal health regimens but under the profound influence of corporate marketing.[23]

Technoscience does not only build vertically; it also moves laterally. A product, procedure, or technique such as gene splicing extends equally to plants, bacteria, nonhuman animals, and humans. For this reason, environmentalists in the U.S. do not stand by idly as biomedicalization and biotechnologization proceed apace. It becomes incumbent upon them to develop positions on innovations that, to iterate Rose, "refigure vital processes" and that enter the marketplace so quickly that regulatory entities often fail to keep pace.[24] If, immediately after World War II, environmentalism targeted nuclear energy as the greatest technological threat, turn-of-the-millennium environmentalism shifts focus to biotechnologies that permeate the quotidian. This shift in U.S. environmentalism in the 1980s and 1990s occurs at the same time as activists are reorienting the movement in two other ways. First, the ambit of its concerns becomes increasingly global, and this, as Ursula K. Heise contends, puts pressure on localism and a parochial ethics of place.[25] Second, the human body is a stage on which environmental risk scenarios play out, resulting in greater attention to how marginalized populations like the urban poor, peoples of color, and indigenous groups bear a disproportionately significant burden of those threats.

Ecosickness is not a comprehensive literary history of these technoscientific developments and their ecological and biomedical ramifications. Rather, its arguments and arc take shape against the background of interventions into planet and body that goad contemporary U.S. writers to reimagine the viability, violability, and value of human and more-than-human life. Ecosickness fiction engages together the histories of environmental and biomedical change evident, for example, in the dark parody of the Biosphere 2 project in *Almanac of the Dead* and the citations of neuroscientific research in *The Echo Maker*. In form and theme, these texts show the conceptual and representational innovations that accompany material alterations to bodies and environments.

One such conceptual update applies to the idea of sickness, which, in this book, is not synonymous with either "disease" or "illness." Arthur Kleinman's and Julia Epstein's definitions of these terms usefully set up the distinction. In their typologies, "disease" implies that there is a biological agent to which medical professionals respond with therapeutic measures. There is a reality to disease in that it suggests the possibility of empirically based diagnosis and treatment, even if the sources of the disease are murky. Thus Laura Bodey incontrovertibly lives with a disease (cancer), whereas Carol White's sensitivities invite skepticism because there is no detectable material agent at their root. She lives with illness rather than disease in Kleinman's and Epstein's schema. *Illness* refers to "the innately human experience of suffering and symptoms,"[26] the "individual's self-perception of a breach of health."[27] Regardless of what blood tests, scans, x-rays, or biopsies might show, illness exists to the extent that someone lives with it and even assumes it as an identity. *Self-perception* decouples person and diagnosis; whatever the content of the diagnosis or treatment might be, the person can determine the form and meaning that illness assumes. Dramas of disease and illness certainly energize ecosickness fiction, but I give preference to *sickness* to emphasize the relational dimension of dysfunction in contemporary narrative. If disease is synonymous with diagnosis and illness with personalized experience, sickness is a relation.[28]

Reading ecosickness fiction, I formulate the following definition of sickness: it is pervasive dysfunction; it cannot be confined to a single system and links up the biomedical, environmental, social, and ethicopolitical; and it shows the imbrication of human and environment. Kleinman gets at these points with his definition of sickness as the "understanding of a disorder in its generic sense across a population in relation to macrosocial (economic, political, institutional) forces."[29] I expand this definition by adding *environmental* and *technological* to this series. Sickness magnetizes these forces and draws them together. In *Infinite Jest*, to take one example, the political and technological adjustments that transform the upper northeast quadrant of the U.S. into a toxic no-man's-land and site for energy production also turn human bodies into sites of phobia, addiction, self-mutilation, and disfigurement. Sickness is a powerful analytic for inquiry precisely because it offers new perspectives on how these macrosocial forces penetrate individual human bodies and how embodied experience might transform these forces in turn.

Intimations of sickness in one's surroundings and in one's own body inflect late twentieth-century existence. Even in utterly remote settled regions such as the high Arctic Circle, inhabitants bear a body burden of manmade carcinogens such as PCBs (polychlorinated biphenyls), and knowing the extent of this burden ruptures the illusion that there are still areas of uncontaminated purity.[30] And just as many Americans obsessively pursue health through exercise and diet regimens, many also proclaim their inevitable sickness fate, their vulnerability to disorder, whether it be asthma and diabetes or depression, anxiety, and attention deficit.[31] Though pervasive and inescapable, sickness is not, therefore, the same for all individuals at all times. Ecosickness fiction attests that geography, wealth, gender, sexuality, race, and ethnicity impact a person's sickness fate, that the criteria for endangerment shift according to these parameters. Sickness is thus both material—above all, bodies are its most sensitive gauge—and subjective—differently lived and demanding representation. For these reasons, sickness is epistemically, ethically, affectively, and representationally disruptive. As the next chapter on AIDS memoirs shows, a sickness that does not arise from environmental contaminants still recalibrates the optics through which a sick person sees, understands, and values her surroundings, and these perceptual and epistemic shifts transform one's sense of obligation to humans and the more-than-human world.

LIFE, ETHICS, AND ACTION

The importance of ecosickness fiction rests on three main contributions to contemporary culture and criticism. First, this literature apprehends somatic and ecological vitality as shared concerns that cannot be isolated from each other, and it simultaneously nuances ecological models of connectedness. Second, it demonstrates the interdependence of narrative strategies and affect and experiments with the ethicopolitical effects of emotional idioms. Finally, this literature expounds how conceptions of agency, ethics, and action mutate under conditions of environmental endangerment and technologization.

Of particular consequence for recent cultural production and the environmental imagination is how this literature apprehends the imbrications of vulnerable bodies in wide-ranging environmental processes. I understand apprehension to be neither mimetic nor passive; rather, as Rob Nixon explains, the term "draws together the domains of perception, emotion, and action."[32] To apprehend in this revised sense, ecosickness narratives depict transformations to life itself under technologization generally and biomedicalization more specifically. The project to reimagine life itself suggests a particular temporality: what we once imagined to be impervious to essential transformation is now undergoing alterations that open up an indeterminate future. Our technologies have made us modern Proteuses, but with a difference that Michel Serres pinpoints: "Through our mastery, we have become so much and so little masters of the Earth that it once again threatens to master us in turn."[33] Authors of ecosickness fiction are not always sympathetic chroniclers of this unprecedented power to change both ourselves—the nature within—and our environments—the nature without. Even as they acknowledge that technoscience is undeniably here to stay, they frequently shun emancipatory techno-optimism and contemplate the consequences of the second mutation that Serres notes: technology mastering us in turn. *Almanac of the Dead* is notable in this regard as it depicts the racist desires that technoscientific practices express; across the ecosickness archive, somatic dysfunction destabilizes techno-optimistic narratives of progress and emancipation.

With these ambitions, writers of ecosickness fiction join social and cultural theorists such as Serres in refiguring life itself. "Life" today yokes embodied humans and nonhuman environments not because they are pristine zones of technological exception but, to the contrary, because they are the primary sites for technological intervention. Ecosickness fiction emphasizes this conception of life as against three received notions of nature: as those parts of the material, organic world that we are not; as an unspoiled static landscape counterposed to civilized and mechanized society; and as the incontrovertible essence of organic matter.[34] Donna Haraway's revision to the concept helpfully unseats these three ideas without then privileging humans as the only makers of the world or announcing a total condition of simulation. She clarifies that "in its scientific embodiments as well as in other forms nature is made, but not entirely by humans; it is a co-construction among humans and nonhumans. This is a very different vision from the postmodernist observation that all the world is denatured and reproduced in images or replicated in copies."[35] Sickness brings this state of affairs powerfully to view, showing the multidirectional relays between different forms of life. Ecological and human biological life are malleable and vulnerable to extrinsic technological interventions and, because of those interventions, are vulnerable to each other as well. Paul Rabinow's prophecies about the turn of the millennium exemplify the trend to conceive of life as malleable due to technological innovations. The anthropologist looks out from the near coincident events of Earth Day 1990, the Rio Earth Summit, and the start of the Human Genome Project and proclaims, "The nineties will be the decade of genetics, immunology and environmentalism—for, clearly, these are the leading vehicles for the infiltration of technoscience, capitalism and culture into what the moderns called 'nature.'"[36] His scare quotes do not signal the end of nature, which Bill McKibben's eponymous polemic declared in 1989; instead, they herald a radical reworking of nature under processes of technological expansion and globalization. This produces a peculiar effect that Bruno Latour concisely captures: "Science, technology, markets, etc. have *amplified*, for at least the last two centuries, not only the *scale* at which humans and nonhumans are connecting with one another in larger and larger assemblies, but also the *intimacy* with which such connections are made."[37] Whether for better or for ill, technology dissolves the human-nonhuman distinction rather than fortify it.

Recognizing this intimacy, ecosickness fiction does not revert to a yearning for purity even as it remains skeptical toward biomedicalization and technoscience. The writers I examine here make palpable the material, conceptual, and affective imbrications of body and environment. In this respect they might seem to elaborate on the ecological perspective of much U.S. environmentalism and nature writing,[38] central to which is a "search for holistic or integrated perception, an emphasis on interdependence and relatedness in nature, and an intense desire to restore man to a place of intimate intercourse with the vast organism that constitutes the earth."[39] Yet they know that the machine is firmly planted in the garden, to borrow from Leo Marx, and explore how interdependence and relatedness are not tantamount to health and harmony.[40] Rather, conflict, risk, discord, and reflexivity are the connotations of interconnectedness for these contemporary environmental artists. With their own take on the reflexivity of technologization, these writers cast relatedness in less auspicious terms than one tends to find in the literature of nature appreciation. In *Infinite Jest*, for example, the U.S.'s innovations in energy production have material and affective consequences that demonstrate the boomerang effect of technologization.[41] While ecosickness fiction rewrites triumphalist scripts of technoscientific development, then, it frequently recodes the affirmative meanings of "intimate intercourse" between body and planet.

In Wallace's novel and across the archive this book assembles, the effects of sickness exceed the capacities of the biomedical and ecological sciences alone to manage them. These effects call out for new stories and tropes and put pressure on extant narrative and emotional templates for apprehending somatic and environmental injury. This brings us to this literature's second contribution to contemporary thought. Just as sickness is the conduit to "environmentality" in the late twentieth century, I argue, affect is the conduit to awareness of the double nature—bodily and planetary—of contemporary sickness and the obligations that sickness entails.[42] Experimenting with the affects of body and earth, contemporary fiction teaches a lesson that environmentalists and social scientists are exploring with increasing fervor: that particular tropes, metaphors, and narrative patterns carry an affective charge that can activate environmental care when empirical studies alone cannot.[43] In the texts that I study here, representational strategies generate narrative affects that produce ethical and political adjustments bearing on the fate of bodily and planetary vitality. These adjustments take place because affect has illuminating and adhesive functions. Because it is transmissible and shows one's vulnerability to the outside world, affect brings to light the connections between self and other that trouble the conviction that humans are autonomous, inviolable beings. Additionally, it provides the conditions for attachment to or detachment from other beings, objects, ideas, and projects. As they "establish salience among the details we observe," affects help us distinguish foreground from background and guide our attention toward or away from matters of concern.[44] The chapters to come investigate how narrative affect lends phenomena relevance, organizes individuals' perceptions, and converts awareness into an ethic. Simply put but not simply answered, which narrative affects motivate action, and which ones obstruct it? Ecosickness fiction contends that, without emotion, environmental and biomedical dilemmas would not enter the orbit of our attention, but it also tells surprising stories about how attention becomes intention.

Chapter 5, on anxiety, delivers one such surprise and demonstrates the salience of affect. In Silko's *Almanac of the Dead*, powerful elites develop technologies such as regenerative medicine and alternative biospheres to control bodily and ecological injury, but these innovations inflict their own social, medical, and environmental damages. Anxiety about such interventions into life itself stokes the novel's fervent call for a revolutionary uprising against technocapitalism and its injustices. Yet anxiety presents problems for political resistance because it suspends action. This does not mark the novel's failure, however. *Almanac of the Dead* is central to *Ecosickness* and to an environmental criticism interested in the mobilizing power of fiction precisely because it cautions that awareness and response are not coterminous. The same emotions that bring us to awareness might orient response in uninvited ways.

I elaborate below on how *Ecosickness* fits into the "affective turn" within cultural studies and offers the environmental humanities new paths for inquiry. For now, I want to emphasize the literariness of affect in this fiction. Rather than argue that a specific narrative strategy necessarily elicits a specific affect in all situations, I show how a text's affective energies depend on the shape of its narrative, its tropological schemes, and the relations between its characters. Each chapter of this study details localized formal elements—such as the toggling between lyrical and discursive narration in Powers's *The Echo Maker*—and their affective correlates—in that case, wonder and paranoia. In addition to text-specific aesthetic strategies, one formal feature shows up in all ecosickness fiction and helps define the mode: the medicalization of space. This literature metaphorizes space in biomedical terms to figure processes of medicalization and biomedicalization. Ecosickness fiction thus announces that biomedicine has penetrated so many domains of contemporary existence that it now colors the ways in which we perceive the physical world. For example, Wallace invents buildings in the shape of organs with circulatory and neurological functions. And this is not the same as the troping of land as body, particularly the female body. Instead, the metaphors that authors of ecosickness invent depend on detailed anatomy and physiology and suggest susceptibility to malfunction rather than beauty and health. Medicalizing space, ecosickness fiction blurs formal distinctions between the traditions of pastoral, wilderness, urban, and nature writing. Sickness ultimately exerts pressure on the form of contemporary fiction and its affective entailments and opens new paths for cultural thought about environmental and somatic endangerment.

One such path leads to energizing people to act on environmental, biomedical, and social injustices. Thus a third central concern of ecosickness fiction is the models of agency that obtain under conditions of global environmental collapse, medical endangerment, and technoscientific expansion. Prima facie, ecosickness fiction might seem consistent with postmodern thought in that it shows the elusiveness of origins, the pervasiveness of disorder, the plasticity of the body and identity, and the potential for macrosocial forces to overpower individuals. Yet this literature also places guarded faith in accountability and action. It certainly recognizes that the foundations for ethical action are always shifting and that intervening in a "world of wounds" might only produce new ones.[45] And yet idleness is not an option. They retain this ideal as they add embodied feeling and aesthetic response to a suite of operations for ethical orientation that includes but deprivileges rationality and empiricism. In these ways Powers, Wallace, and other ecosickness writers push back against postmodern models of detachment. My readings are therefore careful not to mistake ecosickness fiction's skepticism toward one-way linear causality for apathy, indifference, or denialism. Ecosickness fiction does not develop a singular model of agency out of awareness of human-environment imbrication. The following chapters track various ways that contemporary novels and memoirs figure knowledge of human-nonhuman entanglement while grappling with how that knowledge promotes and obstructs environmental ethics and action. As this book argues, affect is central to whether action flourishes or withers. Emotions like the disgust that Wallace's novel develops are key to how a text tips toward or away from care for human and nonhuman others.

A certain strain of posthumanist thought has pursued precisely these questions about how models of human agency transform or deform when we account for "the entire sensorium of other living beings."[46] Ecosickness fiction certainly has affinities with versions of posthumanism that refute human exceptionalism by elucidating the relays between embodied humans and the more-than-human.[47] With these investments, my writers agree with new materialists such as Stacy Alaimo, Karen Barad, and Jane Bennett that humans share embodied being with all manner of life, objects, and forces and that this sharedness necessarily argues against human autonomy and control.[48] New materialists propose that "agentic capacity is now seen as differentially distributed across a wider range of ontological types."[49] These "types" include the industrialized foods, power systems, and stem cells that Bennett explores as well as the "nonhuman creatures, ecological systems, [and] chemical agents" that figure in Alaimo's explicitly ecocritical study.[50] My claims resonate with their models of agency insofar as some ecosickness narratives—*Almanac of the Dead* and *North Enough*, in particular—directly counter human exceptionalism. However, these texts do not entirely express a new materialist sensibility because they do not decouple responsibility and agency. Arguably, Bennett's theory of distributive agency would rankle my writers and many environmental thinkers insofar as it "makes 'responsibility' more a matter of responding to harms than of identifying objects of blame."[51] While ecosickness fiction eschews a simple model of guilt versus innocence regarding environmental and somatic injury, it still maintains that some actors are more accountable than others. This literature thus evaluates conditions for accountability rather than celebrate the indeterminate. As sandboxes for ideas of agency rather than fixed treatises on it, the works in my archive are sometimes inconsistent on individuals' capacity to act. *Ecosickness* does not resolve these inconsistencies but instead elaborates the complexities of narrative affect that produce them.

ECOSICKNESS IN THE FIELD

This book, then, does not only show that sickness is a crucial way that we come to environmental consciousness today. It also establishes that fiction is a laboratory for perceptual and affective changes that can catalyze ethical and political projects. Journalists, activists, and scientists are losing sleep over why their efforts to impart the significance of problems such as climate change have faltered, and they are turning to the arts for other approaches.[52] Narrative and other cultural forms address the lived experience of somatic and environmental threat and the emotional dynamics that environmentalist projects must take into account. To navigate the shifting ground of biomedicine and environmental change, literature takes off from scientific data while humanistically defining the social and ethical dilemmas to which this data points. Experimenting with cultural tropes and narrative affects, ecosickness fiction carries ecological and biomedical concepts into the public sphere, but positions those concepts as inviting quandaries for everyday living and communal response.

Ecosickness fiction produces biomedical or environmental imaginaries through the conventions of postmodernism, apocalypticism, science fiction, memoir, nature writing, the almanac, and domestic drama, among others, without any one strategy dominating. Indeed, one of this book's discoveries is that the representational problems sickness presents cannot be neatly resolved by one mode of storytelling. On this point, *Ecosickness* invites comparison to contiguous studies of the biomedical imagination of contemporary fiction, but diverges from them in crucial ways. On the one hand, my study takes us beyond the domain of science fiction, a biomedical genre that Susan Squier's *Liminal Lives* (2004) investigates. She studies the relation between "technologies of signification" (for her, twentieth-century narrative) and "technologies of recombination" (that is, biotechnology) and argues that the conventions of science fiction bridge these "technologies."[53] Undoubtedly, it is a genre that writers and filmmakers continually adapt to make sense of a technoscientific present and future, but it is not the only genre through which we imagine ecosickness. My study is also in dialogue with Priscilla Wald's invaluable contribution to twentieth-century medical literature, *Contagious* (2008), to which horror, thriller, and action-adventure fictions are central. She delineates an archive of twentieth-century "outbreak narratives," or "epidemiological stories," that draw on public health discourse and transform scientists' and the public's understanding of patterns of infection and treatment.[54] Unlike outbreak narratives, the works under investigation here decouple sickness from microbial sources and invite questions about noncausal, nonempirical ways of knowing. Moreover, ecosickness narratives imagine forms of belonging that are at once more confined—that is, interpersonal and bioregional—and more expansive—that is, transnational—than the "national relatedness" that concerns Wald.[55]

Because no one narrative template will suffice to elaborate the ecological and biomedical transformations taking place today, this book examines a slice of literary production of the past several years. The books of AIDS writer-activists Grover and Wojnarowicz instance the memoir boom that began in the 1990s; when analyzed at all, their texts receive the most scholarly attention from queer theorists. Ecosickness is also fertile terrain for Powers, who represents the proliferation of science- and information-rich novels with humanist investments. Wallace's encyclopedic *Infinite Jest* positions him in a tradition of formal experimentation that includes John Barth, Donald Barthelme, and Thomas Pynchon, among many others, and invites readings informed by postmodernist and media theories. Finally, Piercy's *Woman on the Edge of Time* and Silko's *Almanac of the Dead*, which depict ecosickness as a driver of revolution, figure in the canonization of multiethnic social justice fiction that has occurred since the 1970s.

Contemporary fiction is not the only discursive domain that evidences the intellectual energy somatic and environmental sickness are attracting. Particularly vibrant scholarship on these intersecting phenomena has appeared in environmental history, with Nancy Langston, Gregg Mitman, and Linda Nash using disease as an analytic for explaining how human health, development patterns, and public policies shape each other in the twentieth-century U.S.[56] Literary studies has more to contribute to this conversation than it has henceforth, a dearth that is even more striking within the field of environmentally focused literary criticism, or ecocriticism. Study of bodily and ecosystemic sickness would fit into an expanding thread of ecocriticism that accounts for the embodied, differently lived effects of modernization processes. Lawrence Buell's *Writing for an Endangered World* (2001) made a key contribution to scholarship in this area by delineating "toxic discourse," a literary mode that gives voice to "anxiety arising from perceived threat of environmental hazard due to chemical modification by human agency."[57] Buell's archive reaches back into the nineteenth century and largely consists of works in which pollution, mass urbanization, and industrialization create conditions for environmental endangerment, but developments in technoscience and biomedicine do not receive close scrutiny.

More recent ecocritical scholarship is in line with *Ecosickness* in that as it attempts to understand cultural producers' and activists' efforts to make visible the intimacy between human bodies and the more-than-human world, especially as these are caught up in capitalistic, technological, and geopolitical projects. This book shares with Alaimo's *Bodily Natures* (2010) an emphasis on "a material world that is never merely an external place but always the very substance of our selves and others," but, through sickness, it opens onto other vistas than the feminist intervention that Alaimo achieves.[58] She contends that environmental understanding today must rest on the realization that "we inhabit a corporeality that is never disconnected from our environment," and she looks to twentieth-century culture and activism and feminist and materialist theory to make her case.[59] While Alaimo organizes her study around the varied genres of environmental justice and environmental health activism, I give literary production pride of place and elaborate the formal and affective mechanisms by which fiction expresses human-environment imbrication. With a focus on the genre and style of recent environmental literature, I share common cause with Rob Nixon, whose *Slow Violence and the Environmentalism of the Poor* (2011) details writer-activists' techniques for combatting the often invisible, slowly unfolding violence of military, imperialist, neoliberal, and industrial disasters. He addresses the "varieties of biological citizenship that emerged in the aftermath" of such disasters, but sickness is just one among many metrics of slow violence in Nixon's important analysis.[60] The questions of affect and technologization that motivate *Ecosickness* and its key texts are subordinate to his concern with demarginalizing the global poor and energizing activist collectivities. By focusing on how sickness and the narrative affects that it generates create conceptual traffic between the human and more-than-human realms, my study tells a different story about recent environmental fiction from those that ecocritics Buell, Alaimo, and Nixon offer.

Central to this story is theorizing specific affects that energize or weaken environmental consciousness. Within ecocritical scholarship there has been a call to attend to affect, but this has largely resulted in analysis of the general category of emotional bonds with place and animals rather than of explicit affects.[61] This does not mean that there are no accounts of the emotional entailments of ecological experience in the annals of environmental studies scholarship. The pastoral tradition, so important to environmental thought and representation, undoubtedly has a powerful affective core.[62] Additionally, Yi-fu Tuan's *Topophilia: A Study of Environmental Perception* (1974) and Edward O. Wilson's *Biophilia: The Human Bond with Other Species* (1984) announce the importance of environmental feeling writ large as a guide to place and species relatedness.[63] The centrality of affect to environmentalism can be felt in the polemics that have long inspired environmental artists, ecocritics, and activists—works such as *Silent Spring* and McKibben's *The End of Nature*; they are charged with emotion about the biotic communities and environmentally embedded ways of life that they seek to preserve against impending threats.

Ecosickness dissects the contradictions and ambiguities of affects such as disgust and wonder and argues that they attach, in one direction, to the narrative strategies that elicit them and, in the other direction, to ethical and social dispositions they elicit in turn. The literary objects under investigation here make it clear that the canon of environmental emotion extends beyond pastoral sentimentalism, "expansionist hubris,"[64] and the sublime. As I expand this range, I also establish that even workhorses of environmental affect such as anxiety and wonder do not function as predictably as we might think. The task in the pages to come is to theorize these emotions through recent fiction while also detailing the combined technological, biomedical, and sociocultural mediations that mold them.

Ecocriticism stands to benefit from building the connective tissue between affect studies and environmental and biomedical discourse. More explicit engagement with theories of affect enriches ecocritics' understanding of narrative's influence on how we see, how we know, and how obligations take root. Importantly, this understanding does not come at the expense of close textual analysis, as it is my argument that textual operations set off perceptual, epistemic, and ethical adjustments. In elucidating how recent fiction affiliates somatic and environmental threat through emotion, this book participates in what has been called the affective turn within cultural studies.[65] As Ann Cvetkovich reminds us, this catchphrase suggests that this area of inquiry is novel, and it elides the fact that questions about emotion have long motivated cultural critics.[66] Phenomenology, trauma studies, psychoanalytic theory, queer theory, and microhistories of the social are just some of the domains where emotion has been a key analytic for several decades. It is nearly impossible to unite such a diversity of projects, but two commitments stand out to affiliate accounts of affect in these areas as well as in the cognitive sciences, aesthetics, and political philosophy: determining how objects and events rise to attention in our personal worlds and how attachments, detachments, and commitments form from that attention.[67] Affect is therefore a way of thinking about how the personal becomes social, ethical, and political—how we're stirred to do anything at all or, in the case of ecosickness fiction, anything concerning injured bodies and planet. Arguably, efforts to find a place for emotion in culture and politics become urgent at the turn of the millennium when the commonplace that apathy is the default political mode of Americans is heard everywhere and when the structural transformations of globalization and new media are "delivering" others' stories and emotions to us in unexpected ways.[68]

I will not offer a complete genealogy of affect theory here, as others have already ably done this work.[69] Rather, I want to explain how this book harmonizes with some influential voices in affect studies of late. Sara Ahmed, Charles Altieri, Lauren Berlant, Sianne Ngai, and Eve Kosofsky Sedgwick are some of the brightest guiding lights of *Ecosickness*. These cultural scholars think outside of defeatist narratives of cultural and political apathy and instead attempt to decipher how individuals enter states of what Altieri calls "involvedness."[70] They account for how aesthetic works and social discourses enliven affects that in turn adjust habits of thought and ethicopolitical orientations. Emotions are the instruments with which personal, social, and ethical attachments are built or come undone. They forge a link "between how representative agents are moved and how that being moved positions consciousness to make certain kinds of observations and investments."[71] Ecosickness fiction experiments with affect for precisely this reason. How might emotions make us pay attention to our biomedically and environmentally precarious present? How might they ferry us from awareness to an obligation to respond?

In answering these questions, questions that motivate the following chapters, this family of theorists emphasizes that the awakening of affect and ethical and political commitments does not happen in isolation. Like sickness, emotions are relational; they speak to "the intricate ways we feel our attention and care becoming contoured to other existences."[72] Any aesthetically and politically oriented account of affect is therefore incomplete without an auxiliary account of how feelings constitute or disturb the subject-object distinction. As chapter 4 elaborates most fully, emotions like discord and disgust are disturbing precisely because they, in Ngai's words, "destabilize our sense of the boundary between the psyche and the world, or between subjective and objective reality."[73] Like Ngai, I take the subject-object dynamic as pertaining to readers' relationships to texts as well as to humans' relationship to others, human and more-than-human. Since one of the most profound and yet basic philosophical considerations of environmental thought concerns precisely how different discursive regimes fortify or dismantle human-other boundaries, the environmental humanities need more accounts of how emotions enforce or eradicate these boundaries.

Relays between subject and object are not the only contributing factors to emotional experience and its ethicopolitical upshots. The theorists who inspire me also teach that preexisting evaluative templates mediate affective relations. Many of the affects under investigation here have salience to the extent that they disturb expectations and thus call out the existence of norms and beliefs that might have remained outside of even our own awareness. As Ahmed puts it, "whether something feels good or bad *already* involves a process of reading, in the very attribution of significance."[74] Affects then do not precede norms and beliefs, but they also do not merely follow from them without having their own effect in turn. They have, in Altieri's account, "a complex relation to social context. In one sense [one's] feelings depend on [one's] social station and expectations. But they do not simply reinforce what the society constructs. Rather they indicate fault lines within the prevailing social grammar."[75] My account of discord in AIDS memoirs in chapter 2 bears out this argument. As Jan Zita Grover and David Wojnarowicz move AIDS experience outside the city, discord arises to the extent that their expectations for nature's health and beauty unravel, and this discord irritates ingrained conceptions of nature. Discord then inspires alternative evaluative metrics or the utter refusal of them. Crucially, affect does not just help us choose between available ethical and social possibilities; it sparks the creative work that generates new ones. Altieri announces the importance of the aesthetic thus: "the arts matter for social life primarily because they keep alive the sense that it can be by our objects that we measure our possibilities as subjects—possibilities of response and possibilities of modifying our own priorities."[76] Art carries and energizes affects that allegorize, manage, and sometimes change social and ethical relations. For Altieri and the other theorists named here, art potentiates this form of creativity; aesthetic and discursive projects are therefore at the heart of their theories and of *Ecosickness* as well.

In all of these respects, affect is an aesthetic, intellectual, and political tool, but, as Berlant and others clarify, it cannot be instrumentalized for predictable ends. We therefore find that Berlant's first statements on "cruel optimism" in her eponymous book clarify that optimism itself is "not *inherently* cruel."[77] The volatility of emotions, especially in aesthetic works, is what makes them so relevant to study of environmental relations. This is because, as Karen Thornber details, these relations are almost dizzyingly ambiguous, inconsistent, and sometimes even delusional.[78] Chapter 4 teaches this lesson as it argues that disgust promotes attachment in *Infinite Jest* against our expectations that this affect has distancing effects.

Though indebted to extant affect theory, I do not absorb accounts of emotion wholesale. For example, while Sedgwick undoubtedly broke ground for cultural affect studies in the mid-1990s by fostering affective "reparative critical *practices*" against the hermeneutics of suspicion, my reading of Grover's and Wojnarowicz's memoirs reanimate the suspicion that she hopes to displace.[79] Similarly, Ngai's thesis in *Ugly Feelings* (2005) that "the nature of the sociopolitical itself has changed in a manner that both calls forth and calls upon a new set of feelings . . . perhaps more suited . . . for models of subjectivity, collectivity, and agency not entirely foreseen by past theorists" informs my apperception that ecosickness fiction rearticulates affects that capture experience of somatic and environmental vulnerability and orient response to it.[80] Yet I argue that even as writers like Wallace and Silko mobilize the disgust and anxiety that interests Ngai, they do so in ways that modify her accounts of these emotions.

To Ngai's assertion that the "nature of the sociopolitical itself has changed," *Ecosickness in Contemporary U.S. Fiction* adds that this transformation is lassoed to others regarding biomedicine, technoscience, and the environment. Powerful if sometimes subtle emotions flare up around these actualities with which contemporary artists live. For this reason, the environmental humanities require focused, nuanced accounts of how biomedical and environmental phenomena enter the public sphere through the narrative affects of literary works. Contemporary writers are alert to the representational challenges that sickness presents, and they confront them by innovating ways to tell its stories. Moreover, ecosickness fiction establishes that narrating dysfunction is central to the very definition, experience, and management of it. Ecocriticism must be just as responsive to the affect-rich stories that refract at once thrilling and chilling modifications to body and planet. By addressing the affective potency of our materials, we can speak to the pressing concerns of environmentalists and fellow researchers. This book answers this need through detailed analysis of how recent fiction administers affect not as inherently redemptive—not as a panacea for endangerment—but as a way to direct attention, confirm or undermine ethical stances, and spur or stifle political action. Similarly, my concern is not to specify the "right" narrative affects for articulating an environmental politics but rather to elaborate how unique aspects of the literary ignite feelings that draw planet and body into a shared sphere of concern.

OUTLINE OF THE BOOK

The chapters of this book traverse a range of affects, from the negative (disgust) to the positive (wonder), that recent fiction deploys to imagine pervasive sickness. The book opens with discord through which I outline ideas of nature, health, and beauty that organize perception of earth and soma. Chapter 2 asks: How does knowledge about lived environments change when they are refracted through disease experience? In particular, how does deurbanizing AIDS disturb the metrics by which we value injured bodies and landscapes? In answering these questions, I specify normalized meanings of "nature" that surface in the discourses of ecology and human biology and circulate in AIDS writing and polemics of the 1980s and 1990s. The chapter offers the first account of an unanalyzed mode of writing about HIV/AIDS, the disease that captured medical and artistic attention at millennium's end. Pushing against the metrocentricity of U.S. AIDS literature and scholarship, a body of nonurban AIDS memoirs took the disease out of the city in the 1990s. These stories figure the disconnect between lived experience of AIDS and consensus knowledge about it and show how affect produces irritations that unsettle conceptual categories. Specifically, Wojnarowicz's *Close to the Knives: A Memoir of Disintegration* (1991) and Grover's *North Enough: AIDS and Other Clear-Cuts* (1997) stage unexpected environmental encounters that produce discord. Discord is a powerfully embodied affect that chips away at calcified regimes of thought about nature, in particular the tropological chain that links it with harmony, beauty, and health. Grover's and Wojnarowicz's memoirs model how the irritations of discord generate a productive form of suspicion that confers experiential epistemic authority onto sick bodies enmeshed in their surroundings.

Chapter 2's analysis of the aesthetic, ecological, and moral ideas of nature with which AIDS memoirs wrangle leads into two chapters in which defamiliarizing affects either solidify or dissipate interpersonal and interpersonal connectedness. Through Powers's *The Gold Bug Variations* (1991) and *The Echo Maker* (2006), chapter 3 examines wonder, an affect related to intellection that is essential to inspiring medical and environmental inquiry. Nature writers have turned to wonder to spark ecological awareness as a first step to promoting environmental care. Informed by new scientific paradigms, however, contemporary fiction rethinks the trajectories of wonder. Looking to *The Gold Bug Variations*, I first establish Powers's early vexed considerations of wonder in the context of a burgeoning genetic science. In the later *The Echo Maker*, a plot about neurological damage and research intertwines with one about ecological crisis. On learning that her brother has a rare brain injury, Karin Schluter returns to Nebraska and becomes embroiled in a fight to protect the region's endangered riparian ecosystems and the cranes who nest there. The oscillation between familiarity and strangeness is the ligature between the novel's two narratives as it defines the workings of cognition and of environmental awakening. This oscillation is also the engine of wonder itself. Toggling between a series of binaries, however, the text reveals that wonder can tip over into projection and paranoia, relations that divert energies away from ethical involvement. Ultimately, Powers's ecosickness fiction is a rejoinder to commonplaces about connection and care found in environmental thought.

Powers's vacillations—between the strange and the ordinary, between lyrical and discursive narrative modes, and between ecology and neurology—repeat themselves with different content across ecosickness fiction. In particular, the defamiliarization that these oscillations produce anticipates my study of disgust in chapter 4. *The Echo Maker* leaves readers with the conundrum of whether an ethic of care can take hold without a sense of interconnectedness between self and human and nonhuman others. To this question, *Infinite Jest* (1996) answers, "no." While in Powers an agglomerative affect jams connection, in Wallace disgust, a customarily divisive affect, promotes connection. Wallace's professed ambition to "'author things that both restructure worlds and make living people feel stuff'" assumes that the novel is not only an imagined world but is also an apparatus for reconfiguring the world beyond its pages.[81] The mechanism of this apparatus is affect. This begs the question: how can aversive narrative affects draw a public into concern for deleterious social and material conditions? Chapter 4 finds an answer to this query in the varieties of detachment that circulate through *Infinite Jest*. Wallace's speculative fiction envisions a near-future United States in which the proliferation of media entertainment and consumer goods gives an illusion of freedom that ultimately shades into imprisonment through addiction and solipsism. In this storyworld, detachment is not only a social and ethical problem, however; it is also an environmental one. Psychological and material practices of detachment underpin U.S. policies of environmental manipulation that poison bodies, and it also motivates the novel's major plots and distinguishes its style. Against detachment, the novel deploys disgust as an affective correlate to a medicalized environmental imagination. In detailing the functions of disgust in *Infinite Jest*, the chapter reconciles competing positions within affect theory on whether the feeling is attractive or repulsive. I argue that it is the dual aspect of pulling in and pushing away that makes this affect an unlikely foil to detachment and the social and environmental injustices that it promotes.

Chapters 2 through 4 establish that ecosickness narratives alter standards of perception, representation, and connection through the affects of sickness. Chapter 5 extends the previous chapters' concerns through analysis of anxiety, an affect concomitant to rapidly changing political, technological, and environmental realities. As the inequalities inherent to technoscience came to the fore in the 1970s, women writers crafted speculative fictions to imagine the apocalyptic implications of interventions into bodies and the land. This chapter begins with Marge Piercy's *Woman on the Edge of Time* (1976) to introduce the question whether anxiety about biotechnology can usher in utopian futures. Turning to Silko's *Almanac of the Dead* (1991), I investigate how increasing anxiety about biomedicalization and technological expansion can stymie agency. *Almanac* stages the dispossesseds' attempts to overthrow capitalism and colonialism, which have made the Americas—from Alaska to Argentina—morally, physiologically, and ecologically sick. Against critics who see the novel's revolutionary wishes actualized in therapeutic body-land connectedness, I argue that this template for organizing environmental discourse is at cross-purposes with the novel's apocalypticism. Through scenarios of bio- and ecotechnological horror and medicalizing tropes, body and land become vectors of anxiety. Anxiety is the affect that triggers concern about iniquitous technoscience, but this emotion also neutralizes the capacity to resist its penetration of all domains of existence. Because it compromises human agency, apocalyptic anxiety is at odds with the revolutionary historiography that animates this and other narratives of ecopolitical resistance. Ultimately, *Almanac of the Dead* spans an affective range central to environmental fiction, from harmonious integration with the more-than-human to techno-anxiety, and presents the need for an ecocriticism than can account for this diversity and its consequences for environmental politics.

Taken together, these stories of ecosickness demonstrate that, through affect, environments and bodies become cultural and political. Narrative affects reveal the ecological systems within which we are enmeshed to be more than biological processes; they are conterminous with our own bodies and call out for aesthetic and ethical response. With *fin de millénaire* sickness, new literary modes arise to intertwine the human and the more-than-human, aesthetics and ethics, perception and action. *Ecosickness in Contemporary U.S. Fiction* helps establish recent literature as a necessary complement to empirical modes of understanding the mutating status of life itself. Ultimately, I intend this book to show that sickness is a rich resource for both contemporary cultural expression and environmental thought. While sickness may not be eradicable and will certainly incite more interventions into life, sickness should not provoke fatalistic prognoses. Rather, as the chapters that follow show, narrative sickness opens up a deeply humanistic approach to apprehending our precarious present, an approach that attests at once to the distinctiveness of the literary and to our entanglements in the more-than-human world.

2

AIDS Memoirs Out of the City

DISCORDANT NATURES

PROLOGUE

If the 1970s seemed to confirm a narrative of biomedical progress due to an explosion of vaccines and novel procedures like in vitro fertilization, the next decade brought one of biomedicine's most formidable obstacles. The first cases of HIV/AIDS were diagnosed in 1981, and their intractability shook the confidence of epidemiologists and pharmaceutical researchers. Technoscience could not invent the tools to curb, much less eradicate, this emerging infectious disease in time to prevent massive loss of life. Undaunted, countless scientists attempted to tackle all dimensions of HIV/AIDS, from its pathology and epidemiology to treatment and public health policy. One of these researchers, physician and writer Abraham Verghese, came to AIDS care and research by happenstance; his arrival in the United States as "a rookie doctor" of internal medicine coincided with the first reported cases of HIV.[1] Switching to a specialization in infectious diseases, Verghese treated some of the first AIDS patients in Boston hospitals.

When Verghese left the Northeast for the mountains of eastern Tennessee in 1985, he expected to leave care for people with AIDS (PWAs) behind. No easy feat, as this chapter will tell. The mobility of the virus mimicked his own, and his epidemiological studies and personal writing over the next decade relate how people and microbes migrate between urban and nonurban spaces.[2] *My Own Country: A Doctor's Story* (1994) gives one such account, telling of the doctor's quest to understand his own place and the places of AIDS in the United States. The memoir warmly narrates his discovery of a temporary home in Appalachia, the latest destination in an itinerary that carried him from his birthplace in Kerala, India to Ethiopia, to India once again, and finally to the U.S. The first-person narrative opens with an imaginary road trip on which Verghese is eavesdropping: "Summer, 1985. A young man is driving down from New York to visit his parents in Johnson City, Tennessee. I can hear the radio playing. I can picture his parents waiting, his mother cooking his favorite food, his father pacing.... I can see him driving home along a route that he knows well and that I have traveled many times. He started before dawn.... Three hundred or so miles from home, he feels his chest tighten" (5). This is an anxious homecoming. The son's arrival will shatter the harmony of the hypothetical domestic scene. The transplanted urbanite carries the news that he is sick with AIDS and needs support that his parents may shun or only reluctantly provide. The oscillation between the young man with AIDS and the young doctor that structures this introductory vignette establishes the pattern for the rest of the memoir. *My Own Country* switches between Verghese adjusting to the region's social milieus and physical landscapes and the medical cases that add up to his theory of how AIDS arrives in America's nonurban regions.

In the U.S. the native habitat of HIV/AIDS was and remains the city. Understandably, the city became a metonym for this late twentieth-century medical condition because infections were initially most prevalent in urban gay men living in the coastal zones of New York City, San Francisco, and Miami.[3] Throughout the 1980s and 1990s, AIDS artists codified this emerging infectious disease as metrocentric. Call up the archive of AIDS literature and plots that unfold on city streets come to mind, whether these plots appear on page (Sarah Schulman's *People in Trouble* [1990], Michael Cunningham's *The Hours* [1998], and Thom Gunn's *The Man with Night Sweats* [1992]), stage (Tony Kushner's *Angels in America* [1991] and Jonathan Larson's *Rent* [1994]), or screen (Jonathan Demme's *Philadelphia* [1993] and

Christopher Ashley's *Jeffrey* [1995]). The AIDS canon is understandably metrocentric. In addition to the prevalence argument, we can point to Kath Weston's claim that the "symbolic contrast [between city and country] was central to the organization of many coming-out stories."[4] The city-country contrast became a repeated convention of narratives about homosexual identity formation. Writers created a "sexual geography in which the city represents a beacon of tolerance and gay community, the country a locus of persecution and gay absence," and queer discourse largely followed suit in installing the city as the home of homosexuality.[5]

This chapter does not contest the reasons for AIDS's metrocentricity, but shows that focusing only on the city slights the variety of AIDS writing. An urban lens also obscures the fact that, even as AIDS was affixed to the metropolis, the disease was becoming a signifier of global mobility. HIV/AIDS evolved from epidemic to pandemic because it was never an exclusively place-based phenomenon.[6] *My Own Country*'s opening vignette foregrounds this dimension of the disease's spread, and the text contributes to an underexamined, though rich, collection of AIDS narratives from the 1990s that move depictions of life with AIDS out of the city. This archive includes less prominent works such as Richard Briggs's documentary *AIDS in Rural America* (1990), and novels like Vance Bourjaily's *Old Soldier* (1990), Michael Cunningham's *A Home at the End of the World* (1990), and Geoff Ryman's *Was* (1992), as well as the memoirs that are the focus of this chapter: Verghese's and, more centrally, Jan Zita Grover's *North Enough: AIDS and Other Clear-Cuts* (1997) and David Wojnarowicz's *Close to the Knives: A Memoir of Disintegration* (1991).[7] Shifting the geography of AIDS representation through these works, I join a small group of cultural critics led by Judith Halberstam and Scott Herring who are rethinking the "metronormativity" of queer politics, representation, and criticism.[8] Their scholarship reexamines the ways of life and identity expression available in nonurban places. Yet, even as it redefines the urban-rural distinction, there is scant reflection on how individuals' relation to the environment, rather than their sexuality, changes as they cross geographies. Setting aside the city optic provokes the questions that motivate this chapter's inquiry:

what emotions arise when AIDS experience is transplanted to unfamiliar landscapes in ecosickness narratives? How do these affects produce knowledge about contested environments? And how does this knowledge disturb conventional perceptions of endangered bodies and places?

As the most widely read and critically acclaimed of the three memoirs that this chapter considers, *My Own Country* sets a baseline for pursuing these questions. It establishes the scientific and narrative interest that arose when the paradigmatic disease of the urban century entered nonurban spaces that, on the surface, seemed to offer respite from somatic and environmental injury. Verghese's empirical research and case histories were key to charting the terra incognita of AIDS, and they led a counterwave to the disease's metrocentricity. But there is a disconnect between the doctor's epidemiological and clinical practices and his habits of environmental perception and understanding. My reading in this prologue shows that the genre of the immigrant memoir binds the author in affective handcuffs such that he consistently naturalizes development patterns even as he denaturalizes disease patterns. Verghese sees Appalachian landscapes through an appeasing sentimentality, and this affect informs the normalized ideas of nature that drive the memoir's plot of belonging. It is these ideas that Grover's and Wojnarowicz's memoirs will dismantle. Normalized conventions of nature initially govern their responses to what are ultimately inharmonious environments, but sickness provokes the irritating affect of discord and its ethical, epistemological, and aesthetic payoffs.

The "country" of Verghese's title is multivalent. The memoir is not only a tale of the doctor's adaptation to American life; it is the tale of his adaptation to the landscapes of eastern Tennessee and western Virginia. As a physician at the Mountain Home Veterans' Administration and at East Tennessee State University School of Medicine, Verghese cements his place in these various countries. As HIV/AIDS cases start to emerge, the doctor becomes the region's de facto expert on the disease. Caring for those whom health care networks would otherwise have neglected, he earns locals' trust and respect, and his prestige spreads to the wider medical community when he models a paradigm of rural HIV infection in the U.S.

According to his findings, "infection with HIV in rural Tennessee was largely an *imported* disease. Imported to the country from the city. Imported by native sons who had left long ago and were now returning because of HIV infection" (395).⁹ The lesson: no one can hide from AIDS by leaving the city, a lesson that all nonurban AIDS memoirists communicate.

Yet, for Verghese, rurality does offer an environmental retreat; he "wrestle[s] with the choices: Stay in Boston? Return to Tennessee and raise our baby in a safe, rural, pastoral setting?" (31). His expectation that leaving the city amounts to an escape to a "pastoral" idyll sharply contrasts his consolidating awareness of the region's vulnerability to HIV/AIDS. To put a finer point on the distinction: the environmental health of the countryside is a given while the biological health of its human inhabitants is not. Sentimental ideas of rural nature govern *My Own Country*'s environmental imaginary even as the memoir's biomedical narrative precisely aims to *de*naturalize commonplaces about HIV/AIDS such as the risk factors for infection, how the virus spreads, and what treatment PWAs deserve. As I interpret Verghese's restricted environmental imagination, however, I do not discount his significant contributions to medical research and practice. He champions a sociocultural model of medicine and incisively critiques a health care system that neglects the most vulnerable. Rather than arm himself with technological fixes, Verghese employs other instruments of healing—"hand-holding, family visits, home visits . . . the folksy kind of medicine" (272)—and solicits patients' stories both as a diagnostic tool and "for their own sake" (126).¹⁰

As first-person narrator, Verghese brings the same faculties of keen observation and sensation to bear on both disease and environment and thus invites readers to compare his representation of these two domains that, along with ethnic identity, position him relative to the Appalachian community. The landscape often enters Verghese's view while he is on the move. Driving into a valley in autumn, he notes that "the newly fallen leaves were saffron and ocher in color, their outlines preserved, as if someone had set them carefully by their intact stems to form a mosaic around the bases of the trees. And the trees that soared up from this earth-coat were on fire ... making me feel as if I were driving into a Cezanne [*sic*] painting" (209). Though his attention is meticulous, his perceptions of the more-than-human here and throughout the book endorse three received ideas of nature: as the product of deliberate design ("a mosaic," "a painting"), as harmonious, and as a metric for belonging. The passage suggests that the human eye and the region's flora suit each other according to the intentions of an unspecified agent. Verghese's allusion to European modern art is striking because he more often constructs hackneyed dioramas of rural America. For example, entering "the green, undulating pastureland" of Tennessee, the narrator imagines it coaxing the passerby "to stop and inquire about purchasing the lone house that sat on a hill and all the land around it.... The tobacco allotments in the corner of these estates stood out from the rest of the fields.... In some fields the plants had been harvested and propped up against tobacco stakes in orderly stacks that looked like miniature wigwams" (85). Without irony or critique, this description introduces the three-hundred-year history of indigenous displacement in the South; it places "wigwams" on the very tobacco plantations that may have supplanted native dwellings when white settlements dispossessed tribal peoples. This scene introduces another pattern of displacement that is occurring in 1980s Tennessee: farms are slowly giving way to white-collar homesteads as professionals like Verghese are allured by the "safe, rural, pastoral setting" (31).

In these snapshots of the land, the text is silent on the history of exploitation that underlies Appalachia's apparent harmoniousness, and this silence contrasts Verghese's vociferousness about the sociocultural and historical contingencies that contribute to the region's complex disease landscape. That said, in occasional casual remarks about environmental devastation resulting from industrial practices, the narrative hints that the vistas Verghese admires do in fact have a past. "The lone house on the hill" is enmeshed in an evolving environmental history that contrasts Verghese's sentimental view equating rurality with pastorality. This history includes various forms of mining—underground, strip-mining, and, most recently, mountain top removal—and chemical pollution from the Eastman-Kodak plant and an ammunitions depot in Kingsport, Tennessee. While the text mentions "the acrid scent that came from the plant," the beauty of "the undulating pastureland [that] gave way to hills and deep valleys" effectively neutralizes the ecological threat (85). Even as *My Own Country* charts a new geography of AIDS in demonstrating that nonurban spaces are not refuges from sickness, its reflections on the environment uphold an ingrained dichotomy between "urban war zones" and "rural havens" (22). The natural order remains in balance while the social order fractures and reconstitutes under the pressures of AIDS.

My Own Country's focus on Verghese's quest for an American identity partially accounts for its innocent environmental imaginary as the plot correlates personal adaptation to environmental balance both conceptually and narratively. Verghese suggests that only a timeless, harmonious environment, one free of the sociocultural and historical contingencies that inflect inhabitants' sickness experience, nurtures the belonging that he seeks. Two contiguous reflections on the environments of Mountain Home evince the notion that ecological fit establishes a model for regional and national fit:

> The monotony of the land was broken up by shallow ravines, scattered woods, small rounded peaks, rock formations. The Great Smoky Mountains were a perfect backdrop for it all. The architect's challenge had been to take this land and design clusters of buildings on it of different functions but all of which were to be one homogeneous organism. . . . I couldn't imagine anything other than the French Renaissance buildings of Mountain Home being on this land; they appeared to have grown out of the land just as the magnolia and oak trees had. (294)

Just as the ageless mountains harmonize with the variegated features at their feet, the architect strikes the perfect note with a style that, despite being imported, is as native as the region's flora. Human-made and more-than-human elements meld into "one homogeneous organism." Crucially, Verghese's ability to apprehend the harmony between built and unbuilt environments demonstrates that he too suits his chosen habitat. In a subsequent passage these same oak trees become the ligature that binds

Verghese to nature, region, nation, and planet: "With the flesh of my arms joined to the trees, my feet between its knobby roots, I felt connected to Mountain Home, to my adopted country, to the earth" (299). Verghese detects an ordered, organic pattern within which he slots himself. Even if the order is imposed rather than inherent in the elements themselves, finding a pattern ensures that he also detects his place within it and within the U.S. This adaptation to place, as affect-rich local space and as nation, fulfills the expectation of a memoir of immigration whose title—"my own country"—announces belonging as its telos.

The interaction between disease and environment is unidirectional in *My Own Country*: geographical factors inform Verghese's understanding of the social epidemiology of HIV/AIDS, but AIDS experience does not inflect the environmental complexity and emotional register of his surroundings. In Grover's and Wojnarowicz's memoirs, on the other hand, sickness is a conduit to affective positions that alter environmental perception, aesthetics, and ethics such that the fates of soma and planet necessarily interinvolve. In particular, the authors' nonurban environmental encounters, which take place in the shadow of AIDS, set off the irritating emotion of discord. Simply put, discord is an affect that takes shape when lived experience grates against preestablished expectations. The musical connotations of discord suggest that aesthetic experience nourishes the feeling. This chapter is particularly concerned with how aesthetic response to land and body activates the discord mechanism. An affect that we feel in the gut but that alters conceptual habits, discord in nonurban AIDS memoirs rewires apprehension of North American landscapes and a deadly pandemic. It sets in motion a challenge to the concepts of nature and health that organize these phenomena. *North Enough* and *Close to the Knives*, along with the other authors of ecosickness that this book features, contribute to a new canon of emotions that manages understanding of somatic and planetary sickness by entwining them aesthetically, epistemically, and ethically. Ecosickness renders sentimentalism obsolete and demands that we examine other affects that mold understanding of HIV/AIDS and of the human-nonhuman relation. As chapter 1 relates, bodily injury crucially shapes environmental consciousness at the turn of the millennium. Ecosickness fictions that pair plots of human and environmental endangerment deploy affects that encourage or thwart ethical investment in these dangers. Representing vulnerabilities of planet and self, *North Enough* and *Close to the Knives* avoid the teleological plot that *My Own Country* features and expand the possibilities of the memoir genre.

What happens when sick bodies enter new spaces? This simple question launches the readings that follow and charts a new trajectory for study of AIDS writing that has centered on how individual and group identities dissolve or consolidate around sickness. Surveying the rhetorics of nature that surface in AIDS discourse of the 1980s and 1990s, at a time when nature-skepticism reigns in cultural and ecological theory alike, this chapter begins with the persistence of a conceptual chain linking nature, health, and beauty.[11] As Grover's and Wojnarowicz's memoirs leave behind the city as the "privileged setting" of AIDS culture, they, on the one hand, illustrate that this conceptual chain binds them and, on the other hand, show that disturbing affects like discord can be tools for dismantling it.[12] Discord confers on these writers an epistemic agency that is based in the body. The memoirists heed the destabilization of discord and reevaluate the conceptions of nature, health, and beauty that first come readily to mind. Essential to this agency is the efficacious form of suspicion that discord catalyzes. While suspicion has come under attack from cultural critics of late, *North Enough* and *Close to the Knives* show suspicion's detractors that this stance can generate new ethical and epistemic orientations rather than stifle them. Discord and the suspicious stance it sparks do not only matter to memoirists' private aesthetics, epistemology, and ethics, however; the affect has implications for oppositional politics as well. Akin to the contemporaneous environmental justice and AIDS health movements with which this chapter concludes, Grover's and Wojnarowicz's projects validate experiential authority as they contest normative knowledge of sick bodies and environments.

CONTESTED NATURES

Shifting the geography of AIDS furthers Susan Sontag's project to expose how representational strategies either bind us to habits of thought about disease or liberate us from them. Dismantling master tropes is one of the key weapons in the "struggle for representational ownership" of early AIDS in America.[13] A critical focus on cultural politics was integral to this "struggle." Artists, scholars, and public thinkers used metaphor, imagery, slogan, and narrative to erode the discursive regimes governing sex and health practices. Underpinning these regimes was a sense of the naturalness of particular bodies, desires, and actions. Indeed, in an acerbic observation, Grover notes that "nature" is on the tip of everyone's tongues in AIDS debates: "Nature, it turns out, is the last refuge of the scoundrel. . . . When all else fails the mainstream commentator on AIDS, s/he still has at [sic] disposal the seemingly irrefutable argument that x or y is unnatural and therefore deserves or inevitably brings on Nature's revenge for (name your vice): promiscuity, homosexuality, non-reproductive (usually anal) sex, drug-use."[14] In Grover's estimation, nature is cloaked in a moralistic mantle and the concept then justifies the "scoundrel's" positions on AIDS and queer life. Evangelist Jerry Falwell is a likely referent for the damning sobriquet that Grover uses, as he was a mouthpiece for religious fundamentalist disapprobation of so-called crimes against nature and those who "perpetrate" them.[15] For Falwell and politicians like Jesse Helms, nature underwrites a model of AIDS "as symptomatic of a breakdown of moral hierarchies, order and authority, thus requiring a primarily moral solution, in order to 'save' the Holy or the Pure."[16] The disease both provides proof of that moral order and is an appropriate punishment for violating it. Seemingly irrefutable, as Grover notes, the model is appealing in its simplicity: a breakdown in the body's immune system, the mechanism that maintains internal order, signals a breakdown in God's ordained external order. AIDS activists and scholars direct most of their energies toward contesting this meaning of "nature," but the concept fills many other roles in AIDS discourse as well. In addition to being a metric for morality, three secular uses of nature ring out that I outline here: nature as geophysical force, as human biological and ecological substrate, and as discursive.

Stephen Jay Gould fixes the first of these uses in a 1987 *New York Times Magazine* essay. He strips AIDS of its "moral meaning" by calling it a "virulent version of an ordinary natural phenomenon."[17] Gould's is a morally neutral but still "endorsing" sense of nature that denounces the vitriol of bigoted fundamentalists. Akin to earthquakes and floods, AIDS reminds technologically dazzled Americans that "we have not canceled our bond to nature." This geophysical understanding of AIDS's naturalness raises fewer hackles than those denotations based in erroneous conceptions of human biology. In a foundational essay on AIDS as an "epidemic of signification," Paula Treichler cites a 1985 *Discover* article that surveyed the epidemiological research to date and perpetuated the wild falsehood that women who had never had anal sex were invulnerable to HIV.[18] John Langone, the article's author, "suggests that the virus enters the bloodstream by way of the 'vulnerable anus' and the 'fragile urethra;' in contrast, the 'rugged vagina' (built to be abused by such blunt instruments as penises and small babies) provides too tough a barrier for the AIDS virus to penetrate."[19] This is just one of myriad misconceptions that guided research agendas and public health procedures and, in many cases, slowed prevention and treatment policies that could have made the syndrome more livable. Common to such falsehoods about human anatomy and even ethology are unexamined assumptions about how the natural body, which is also the healthy body, functions. In other words, demagogues, scientists, and journalists alike legitimate some bodies and practices against others by measuring them against an ill-informed biology. In Catherine Waldby's words, they "exploit the nature/culture binary in ways which create hierarchies of pathology along the lines of sexual identity. [They] use unexamined and naturalized notions of sexual identity, of male and female, and of heterosexual and homosexual, to 'explain' and thus to control the transmission of natural virulence into culture."[20] This paradigm in turn informs explanations all the way down, to "viruses, cells, genes and immunological communications."[21]

Gay writer and activist Gabriel Rotello engages in another naturalizing project that gets its impetus from the ecological rather than the biomedical sciences. Advocating behavioral over technological solutions to AIDS transmission, Rotello argues that promiscuous gay men have created an "ecological opportunity"—a niche—within which HIV could thrive.[22] "Unrestrained multipartnerism" creates the "sexual ecology" of his book's title; it is this expression of sexuality rather than the virulence of the microbe itself that produces a deadly epidemic, or "ecological catastrophe".[23] Rotello's understanding of ecological science and of deep ecology is limited and often incorrect, and his applications of ecology and environmentalism to male homosexuality are specious at best and moralizing at worst. I emphasize one point here. Even as Rotello admits that the sexual ecology of American homosexual men is a "cultural system," he appeals to ecology because it seems to offer him a culture-free ground from which to critique gay men's failure to prioritize the community's general health over any other values and pleasures.[24] Health and nature line up as metrics for evaluating the right- or wrongness of sex practices in the "ecosystem" of gay men.[25]

At the same time as religious leaders like Falwell, scientists like Gould, and writers like Rotello were deploying these sometimes competing ideas of nature, poststructuralist cultural theory was jettisoning naturalism of all kinds in favor of discursivity. Antinaturalism predates the HIV/AIDS crisis; we can locate one of its sources in Michel Foucault's early work, especially *The History of Sexuality* (1976), and French feminist theory of the 1970s. But it reaches its heyday in the 1980s, when culturalism and constructionism become the lingua franca of humanistic scholarship and inform feminist, queer, and other activisms. Biology is not deterministic, the story goes. Rather, the historically and socially contingent discursive domain shapes even those beliefs and experiences that appear most encoded in our DNA. No appeal to a prediscursive order, whether it is divine or biological, can warrant claims about sexuality, embodied life, or identity and its expressions. Antinaturalism can be a powerful rejoinder to the reductionism that underpins the injustices of racism, sexism, homophobia, and ableism.[26] As philosopher Kate Soper notes, however, "those denying the 'naturality' of sex and the body . . . have purchased the 'freedom' of human sexual practice from any dependency on or determination by biology only at the cost of sacrificing all explanatory and prescriptive force."[27] In brief, antinaturalism can undermine the authority of embodied experience. When the body is only conceived as a repository for social practices or site of resistance to them, the body becomes a shakier foundation on which to build a politics or ethics because its materiality has diminished. It's for this tactical reason that, according to Foucault's history, "homosexuality began to speak in its own behalf [in the nineteenth century], to demand that its legitimacy or 'naturality' be acknowledged, often . . . using the same categories by which it was medically disqualified."[28]

AIDS in America thus stretches the concept of nature into often polarized but sometimes unexpectedly compatible usages at this historical juncture. It invites opposed nature-endorsing positions as well as the nature-skepticism of the constructionists. Soper negotiates these positions, explaining that the body "is not an artificial construct but a subject-object, a being that is the source and site of its own experience of itself as entity".[29] It is both undeniably material and experienced only through the socially shaped self. Her middle path between naturalism and antinaturalism informs my conclusions about the experiential authority of discordant AIDS bodies at chapter's end, but my point here is that environmental and sexual politics meet and sometimes compete on the conceptual battlefield over nature. Above all, the uses of nature swirling in AIDS discourse are important to the memoirs to which I turn shortly because, even as these texts expound on the moralistic, geophysical, and human and eco-biological conceptions of nature, they underline its ecological and topographical meanings. That is to say that nature in *North Enough* and *Close to the Knives* is the material stuff of the world: trees, mountains, animals, flowers, and wetlands. They retain the facticity and political power of materiality and embodiment while chipping away at "nature"'s cultural encrustations.

Confirming the materiality of the more-than-human while reconceiving its meaning, Grover and Wojnarowicz speak to shifts occurring within the ecological sciences in the second half of the twentieth century. Environmental historian Daniel Botkin chronicles ecologists' and environmental activists' robust protest against the normalization of nature beginning in the 1950s. In this period, discordance began to drown out harmony in the symphony of thought about ecology and nature more broadly. And yet, in subsequent decades and into the present, the sacred trinity of harmony, balance, and order continued to endure as the master trope that molds apperception of environmental and human functioning alike. For both laypeople and ecological experts, a default expectation is that ecosystems tend toward balance. The words of nineteenth-century conservationist George Perkins Marsh in *Man and Nature* (1864) represent and inform this perspective. He sees "man's ignorant disregard of the laws of nature" in the forests "blasted with sterility and physical decrepitude" and advocates the "restoration of disturbed harmonies."[30] Over a hundred years later, adherence to an ecological model predicated on ideals of harmony and balance persists; the conviction that "the biological world is one of marvelous order" still warrants environmentalist rhetoric and policy when Botkin writes in the 1990s.[31]

Basing their ethical and policy arguments on nature's harmoniousness, environmental activists, writers, and managers uphold the longstanding intellectual tradition of seeing in nature an order either established by a divine progenitor or resulting from evolutionary progress.[32] Left out is an alternative stream of thought with equally ancient roots that sees nature as "'by its very essence discordant, created from the simultaneous movements of many tones . . . leading not to a simple melody but to a symphony at some times harsh and at some times pleasing.'"[33] In this model, ecological systems are not configured to maximize and harmonize the diversity of life; thus, the telos of harmony does not guide environmental policy on issues ranging from forest management to genetic modification.[34] We must discard this outmoded measure of an environment's health because, in short, "the old idea of a static landscape, like a single musical chord sounded forever . . . never existed except in our imagination."[35] Botkin's counsel to ecologists and environmental professionals chimes with Dana Phillips's injunction to nature writers and ecocritics. Phillips charges that the latter adopt outmoded ecological models as tropes in order to promote ecological belonging and to contest social constructionism. With these aims, the argument runs, wrongheaded environmental writers and critics have cleaved to the antiquated metric of harmony and valorized nature writing as a transparent window onto nature.[36]

Of course, ecological theories continue to evolve; the field is as stochastic as the theories that supplanted the classical model against which these scholars write. Botkin's and Phillips's protests are very much of their time: for Botkin, a moment when cultural theories that denaturalized human attributes, desires, and practices entered the mainstream and, for Phillips, the literary and ecocritical pushback against this intellectual trend. What I emphasize here is that a large cohort of scientists, cultural producers, and laypeople all tend to seek harmony in nature no matter what theoretical paradigm reigns, and this tendency perdures, I claim, because of the penchant to homologize human biological and ecobiological systems. In this homology the organic realm as a whole assumes the form of a subset of it, the human body. Because the body in all of its complexity (most often) functions so elegantly, it must be a balanced system that interacts with a formally parallel system. The body is both metonym for and proof of an ordered nature. Admittedly tautological, this formal parallelism between body and environment joins up with an operational parallelism in views of nature. Failure—that is, disease—isn't normal. It isn't "natural." The right body is the healthy body.[37] To the extent that body and ecosystem are analogues in this way, nature and health link up in a conceptual chain that, as Verghese's environmental sentimentalism suggests, is affectively satisfying.[38]

It is also aesthetically satisfying. Before elaborating how this chain entangles Grover and Wojnarowicz, I hook one more term onto it. The formal homology between human bodies and more-than-human nature does not only promote the idea of planetary harmony; it also dictates aesthetic orientations toward nature. Medical, ecological, and literary discourses are so easily blinkered by harmoniousness because it accords with a paradigm of aesthetic value operant since at least the Hellenic era. Whether in the grip of pastoral sentimentalism, sublime awe, or scientific wonder, and whether peering at the single flower, the sweeping prospect, or the double helix, the observer of the more-than-human world delights in the supposed symmetries that greet her. Universalized but ultimately contingent aesthetic principles then mold response to environmental and somatic dysfunction.[39]

At the end of the millennium, ecology, poststructuralism, fundamentalism, and cultural production share a field of contests over nature's meaning, value, and tropological power. In the works to which I turn now, AIDS memoirists carry the conceptual baggage of "nature" as they relocate experience of the disease out of the city. *North Enough* and *Close to the Knives* demonstrate that the bonds connecting beauty, harmonious nature, and healthy body are persistent and potent. But they also show that, when perceived as inextricable, somatic and environmental sickness incite affective adjustments that enable them to discharge that burden. Analyzing how these memoirs depict bodily sickness and environmental experience, we see why normalized conceptualizations of nature persevere and how they burst when under the pressure of discord.

NORTH ENOUGH'S "DIFFICULT BEAUTIES"

In the early 1990s Grover leaves the Castro district of San Francisco, where she nursed and counseled HIV/AIDS patients for the Shanti Project and local hospitals. Her time as a caregiver and patient advocate coincided with a time "when some workers... still refused to bring trays to the bedsides of people hospitalized with AIDS."[40] The imprints of ignorance, fear, loss, anger, and futility are deep, despite her commitment. Seeking a "geographic cure" for her emotional and physical exhaustion, she travels to her mother's natal grounds in the north woods of Minnesota (6). "Move on, the fantasy ran, and your problems would be different—either that, or they would simply go away" (4).[41] Grover responds to what Ann Cvetkovich identifies as the "'secondary traumatic stress'" of the AIDS crisis: that is, "the forms of trauma experienced by those who work with trauma survivors," like the many lesbian nurses and counselors caring for HIV/AIDS patients and friends whose injury from sickness experience goes unacknowledged.[42] If moving from city to woods is to "cure" this stress, Grover must assiduously attend to her present surroundings at the same time as she gives memory space and time to dilate.

Sheltering both recollection and observation and leaping between years and geographies, *North Enough* has a rich texture. Memories of lost patient-friends cohabit with Grover's consolidating impressions of the bioregion straddling the U.S.-Canada border. Published in 1997, the book belongs to the explosion of first-person life writing that coincided with the spread of HIV/AIDS and that continues into the twenty-first century. The memoir genre has been hospitable to accounts of personal trauma, loss, injury, and disease, a fact that has earned it the often dismissive sobriquet *misery lit*.[43] To the extent that Grover's text focuses on disease and dying, it fits neatly into the category. Yet, in particularizing her environments, Grover stretches the genre. As Cvetkovich notes, the text traverses the boundaries between activism and caregiving and between self and other. She pushes the "I" into the background and "expresses her feelings in feelings for others."[44] Crucially, these "others" include not only friends and PWAs for whom she cared but also the northwoods landscape. *North Enough* satisfies readers' generic expectations with its predominantly first-person narration and confessional tone, both of which appear in the first sentence, "I did not move to Minnesota for the north woods," but the north woods also subordinate the ego (3). The "I" goes silent for stretches of several pages; it becomes embedded in a collective "we" or in an unfocalized omniscience that matter-of-factly delivers information about her new environs: "The Minnesota and Wisconsin cutovers are northern counties that were logged over several times between 1860 and 1920" (14). In this way, *North Enough* is polyphonic: it carries Grover's voice as the locus of secondary sickness experience in concert with the voice of an ecosystem and patient-friends that have suffered primary injury.

In addition to leading her away from exclusively first-person narration, Grover's naturalist program lends the text a Thoreauvian flavor. Claiming to want "a still point from which to reflect on losses [she] had yet to claim," she is in fact a restless wanderer who imbibes her surroundings (75).[45] Like the Concord native, she cultivates a "habit of attention . . . a true sauntering of the eye" as she wanders the woods, and, once home, transforms her observations into prose.[46] She might take her cue from hermetic nature writing, but her training as a historian also places her in the company of those like Aldo Leopold and William Cronon who adjust environmental perception by historicizing the landscape. Grover finds everywhere the evidence of forest and freshwater management. Her environmental vision incorporates not only the "radiant diversity" of native species but also "perverse" necrophagous bogs and, most importantly, the clear-cuts of the book's subtitle (94, 139). Through her historicized naturalism, Grover "learn[s] local patterns but stand[s] at a slight remove from them, like the chronically ill, the very poor, and the otherwise disenfranchised" (161). Here, and throughout the memoir, Grover suggests that, while she is not HIV positive, the virus has infected her such that she sees with the eyes of the sick. Specifically, living and working with PWAs has had a dual effect on her perceptual and conceptual procedures. On the one hand, this experience shrouds everything in a mantle of decay. On the other hand, bringing AIDS to the woods helps her shake off naturalized norms about nature that govern how she apprehends her chosen home. Juxtaposing "close, wondering observation" and a record of HIV/AIDS, Grover's ecosickness narrative identifies how somatic sickness alters habits of perceiving the nonhuman environment (121).

Grover first enters the north woods as winter is turning to spring, a time when "the birch, aspen, and tamarack [are] skinned of their needles and leaves. I thought they looked diseased. But in those days I saw disease everywhere" (3). The "scabby forest lacks any beauty I can understand," she remarks, but it shows the epiphenomena of dying and decay: "Scoriatic bark, rheumatoid branches, the torn flags of last season's leaves" (18).[47] Her metaphors powerfully convey the slow burn of her burnout from AIDS work, and they do so by medicalizing the landscape. In this respect, *North Enough* shares a strategy with other ecosickness narratives: it figures the nonhuman environment through tropes of the body, a practice reaching back at least to the early modern period, but updates this practice by infusing those tropes with anatomical detail in a medical register.

Several months later, tints of "waste" and "desolation" continue to paint Grover's travels northward (152). Visiting the subarctic zones of Manitoba, Canada, she chastises herself for her stunted vision and continued failure to be "sympathetic to [the woods'] homeliness" (18). Ugly, flat, and expansive, the landscape elicits insecurity: "My initial reaction, though I know that at most I can be no more than three blocks in any direction from my rented room, is one of panic.... And that, of course, is precisely the reaction I want to resist, the involuntary reaction.... Few Canadian Europeans see variety or plenitude in tundra, taiga, or muskeg. Rather, they see these as waste places of intolerable solitude and desolation" (152). Grover's myopia irritates her. It's as if two landscapes spread before her. Hobbled by an "involuntary reaction," she only sees one that is emptied of vitality while that which is filled with diverse marvels hovers outside her field of vision. The expectation that nonhuman nature will express health and beauty and will be curative initially determines Grover's primary, unreflective reaction. She demonstrates that the conceptual chain that equates nature, health, and beauty encircles her: she leans toward a deceptive tropological and environmental comfort zone. Like the Canadian Europeans, she prefers to derive "easy" pleasure and feelings of security and belonging from her environs (115). Grover's ready slippage into naturalized codifications of nature demonstrates Sara Ahmed's claim that "normativity is comfortable for those who can inhabit it. The word 'comfort' suggests well-being and satisfaction, but it also suggests an ease and easiness."[48] Molded by the devastation of sickness, however, Grover's new standard is dis-ease and discomfort. She fixes her eyes on the "difficult beauties" of so-called waste places, beauties that arise when, with the eyes of AIDS, she looks on a landscape transformed by centuries of logging and bog draining (6, 152).[49]

Dismantling standards of aligning nature with health and beauty, *North Enough* fits the project of queer theory and regionalist writing alike. Akin to the women's regionalist writing that Judith Fetterley and Marjorie Pryse describe as a form of "'queer' fiction," the memoir scrutinizes extant measures of value and "challenges a 'wide field' of 'regimes of the normal.'"[50] As a kind of regionalist, Grover "takes up the question of the 'peculiar' and the particular rather than what is 'universal'" and develops novel heuristics for valuing and belonging to place.[51] Though in line with this version of literary regionalism, Grover departs from it in making sickness, both somatic and environmental, the key device for recalibrating conventional ways of apprehending a place. Her experiential AIDS knowledge is the wedge that begins to separate health and beauty from nature. Like people with advanced AIDS, the clear-cut and drained north woods are superficially damaged, even ugly, but they harbor depths of beauty and suggest other vitalities. These vitalities emerge in one of the text's early descriptions of the diseased body. Looking on her ill friend, Perry, Grover figures the diseased body in zoological, botanical, and aesthetic terms as she zooms from surface to depth. (We can think of the botanization of the body as the inverse of medicalizing the landscape.) She observes his "edematous face, a lesioned and smelly body," but her gaze penetrates this surface just as "he sunk deeper and deeper in his bed, as if growing roots, extruding layers like a clam, blanket upon blanket, until he was little more than a finely carved mask with the high white nose of the dead, surrounded by petals of bedding" (20, 13). Grover recodes sickness as a naturalized beauty characterized by growth and proliferation before exalting Perry into an art object posed on a flowerbed.

I comment below on Grover's questionable practice of aestheticizing sickness, but first pause on how naturalizing and beautifying the diseased body feeds back into environmental relation. Grover's "habit of attention" takes in evolving social, economic, and political configurations. She entertains the "sentimental response" to landscape but, unlike Verghese, does not let it linger (21). When she looks on the "natural, natural, all so natural" forest, she interrupts herself—"And yet"—as she identifies the cicatrices of past injuries and extant destruction that make her question whether "natural" is the right adjective. Mustn't nature, she implies, be a "rural haven" or "perfect backdrop" as it is for Verghese?[52] Detecting the mutations that have turned bay into bog, Grover interrogates her templates for categorizing her environments as either natural or degraded: "What model—too inappropriate, too human—do I use when embracing this landscape and calling it damaged, imperfect?" (23). Though all "too human," Grover's model holds in suspension the seemingly incompatible: affection and recognition of injury. One way in which she achieves this balance is through language; she assigns semantic plenitude to apparent wastelands. Language bridges the memoir's naturalist and ethical projects in that the naturalist's practice of multiplying topographical descriptors amounts to a plea to recognize the ecological richness of even sick environments. In effect, language unearths reserves of environmental value that centuries of equating bogs with waste places have buried. A bog has a "scant beauty," yet a robust vocabulary accrues in indirect proportion to this aesthetic slimness (138). Over three lines, Grover excavates synonyms for "bog": "*string, flark, carr, schlenke, strangmoor, palsa, lag, moat, moor, muskeg*" and more (138). The harsh, Germanic words evoke the bog's supposed ugliness and abuse, but the passage releases new histories as it accumulates synonyms. Like "ancient forests," bog ecosystems accommodate "inhabitants of many ages, varieties, and a rich, rich understory," a word that has ecological and historiographical registers (75). Compared to the forests' ecological and semantic density, even San Francisco's vibrant Castro neighborhood feels like a "monoculture" (75). Linguistic fecundity contrasts the singular and yet conceptually overburdened word *nature*. Cognate with universalized concepts of beauty and health, "nature" is dense to the point of petrification. Grover's performed defense of diversity then corrodes these hardened conceptions.

Sickness might cause Grover to "see disease everywhere" (3), but this is only one of AIDS's effects on her perceptual habits. Because it drives her to identify vulnerability and "measureless damage" as the very essence of the natural, AIDS experience also retools her environmental aesthetic and ethic (21). Like the sick body, "landscape presents itself as an epistemological puzzle. Can we understand it by recurring to what it once was?" (21). How can we determine the standard for what "once was"? When it comes to valuing the present and future states of earth and soma, too often that standard is an impossible beauty and health, "a more than century-old romantic American tradition that equates the nonhuman with the pure and virtuous" (115). Grover tags as specifically American what Botkin identifies as a more broadly Western tradition, one informing medical and ecological directives to restore all landscapes to an ideal state that "never existed except in our imagination."[53] *North Enough* develops alternative perceptual, epistemic, and ethical templates by sloughing off this tradition and instead engaging in two projects based in observation of the as-is. First, Grover refuses to compartmentalize "human" and "nature," and, second, she promotes a thanatological aesthetic.

The conceptual boundaries separating bodies from environments erode when Grover apprehends that they cycle through homologous states of sickness and fecundity. The memoir employs the guidebook and the landfill as objects through which to articulate how aesthetic categories cement a troublesome human-nature dualism and to imagine the imbrication of human and more-than-human. You won't find landfills in any guidebook's index, Grover reminds the reader. Guidebooks depict "a peculiarly prehuman (prelapsarian?) world free of high-tension wires, condo developments, country roads, and dump sites. A world so absent of human intervention is not only teleologically suspect; it is ecologically absurd. No northwoods habitats remain unaltered; no current forestscape is much more 'natural' than a landfill" (101). To remain practicable, the nature concept must shelter such "blemishes" as landfills, because what are on one level desecrations are, on another, reparations for other injustices. Grover informs readers that, "for our fellow omnivores, [landfills] have been practical blessings, one of the few exchanges we've offered for appropriating so much of their original habitat" (102). They are supermarkets for bears and less charismatic species like crows, vultures, and gulls whose habitats have disappeared with encroaching development. Grover goes against the tide of antilitter environmental campaigns of the 1970s and 1980s and redeems trash as reparative and compensatory rather than contaminating. She observes that birds repurpose discarded condoms and beer cartons into shelters, using them as liners for nests on their migrations through the north woods. Identifying the interdependence of humans and such animals, the memoir answers historian William Cronon's challenge "to stop thinking of [the natural and the human-altered] according to a set of bipolar moral scales in which the human and the nonhuman, the unnatural and the natural, the fallen and the unfallen, serve as our conceptual map for understanding and valuing the world."[54] Grover, too, historicizes her environs, not in search of a lost but possibly recoverable past, but in search of adaptations to encroachment and loss. Seeing versions of naturalness in the human-altered landscape enables us to apprehend longstanding circuits of dependence and new ones that our actions set in motion. Grover understands the north woods not simply as degraded and injured but as integral to a calculus of damage and compensation. This new understanding of the place shapes an environmental ethic of close observation, historical and linguistic excavation, and appreciation of vitality in injury.

For the calculus of loss and recompense to figure, Grover must cast out typically sentimental affects like nostalgia that, along with the denigration of the ugly, preclude us from discovering resilience in the face of damage. Ecological and somatic injury then elicits two not entirely compatible responses from Grover. On the one hand, she deplores humans' encroachment on other species and the tendency to decimate "radiant diversity" rather than observe and protect it (94). On the other hand, *North Enough* feeds off of the death that biological processes, human industry, and disease produce. Like the bog plants it celebrates, the memoir "liv[es] atop the many strata of [its] dead" (139). Grover confesses as much when she declares, "If I choose to find meaning in any of this, I must remember it is my meaning, just as the comfort I drag from friends' deaths . . . is for and by myself" (21). A thanatological aesthetic in which sickness and injury are redeemed as the quintessence of nature confers value on decay. This aesthetic is an environmental and medical ethic as well, one that rejects a premise of many land reclamation projects and medical therapies: the idea that everything can and must be restored to an impossible ideal of health and beauty. Instead, Grover's aesthetic promotes an ethics of resilience and adaptation. Using sickness as something "to-think-with" rather than to turn thought away from, Grover welcomes affects such as the discord to which I turn below that scramble the equation of health, beauty, and nature (95). She in turn rearranges the lines of contest over these sites, refusing to pit "evil" wood products companies against "holy" tree huggers. Rather, the political waters are far murkier as she pours in the ecological, aesthetic, and, yes, writerly benefits of environmental and somatic sickness.

Encounters with nature—as physical reality and as concept—are necessarily mediated by Grover's AIDS experience. Similarly, her aesthetic and ethical orientations toward devastated bodies and environments arise only when she leaves the city. In the north woods the meanings of beauty and nature that sickness has configured come into conflict with the inherited ideas of nature that—and this is crucial—she herself first expresses, but then casts off. Assigning aesthetic and ecological value to sickness, both human and nonhuman, *North Enough* draws out but then dissolves the conceptual equation of nature, beauty, and health. The compromised but still vital Minnesota woods, a space seemingly foreign to AIDS, produce friction between the tropological chain to which Grover still adheres and her lived experience of sick bodies and ecosystems. The friction that ecosickness elicits is the irritating affect of discord.

Before continuing to elaborate this discord in Wojnarowicz's nonurban AIDS memoir, I pause on Grover's practice of aestheticizing injury and death to promote an ethic of adaptation. Catriona Mortimer-Sandilands approvingly describes the ethic that emerges in *North Enough* as a form of care that "involves making choices to observe and act in the world mindfully: to care for the world as it is, as we have contributed to making it, rather than as we would like it to be ideally."[55] On this point that the memoir rejects ideals, Mortimer-Sandilands and I agree. But the memoir's ethical and aesthetic positions are not so easy to swallow. Encompassing the sick, blemished, and irremediably altered, Grover's nature embraces and even exalts an as-is that includes "Styrofoam and clear-cuts" (123). Environmental blights to most, "they as well as the centuries-old stands of white pine can be turned to holy purposes. Disasters are historic, local, specific: redemptions are as well. . . . The clear-cuts are filled with north woods in which to worship and work for miracles" (123). No environment is forsaken. Grover's position on ecological salvation is appealing in its hopefulness, as Mortimer-Sandiland's response indicates. But can we be taken with the beauty and benefits of destruction and still direct our energies at eliminating the sources of that destruction? This question is germane to Grover's environmentalist and AIDS commitments alike. While the ecosystems she traverses might be put to new "holy purposes," PWAs are not so fortunate. In a sense, they are like the bog plants at which she marvels, the decaying but beautiful matter from which new life grows. In this case, that life is her own and that of her memoir. Redemption comes to the assiduous observer enriched by the lessons of environmental and somatic sickness, not to the dying and the dead. While it may be difficult to embrace Grover's ethics of redemption of the damaged, the affective relation from which it emerges—that is, discord—does not have a predetermined, singular outcome, as Wojnarowicz's memoir shows.

THE "CON" IN *CLOSE TO THE KNIVES*

If redemption is available but problematic in *North Enough*, it is elusive in Wojnarowicz's "memoir of disintegration." *Close to the Knives* contrasts Grover's naturalist recollections with its more radical form and furious critique. Yet it too challenges its author's inherited preconceptions of nature and posits discord as the affective correlate to the perceptual, epistemic, and aesthetic adjustments that occur when AIDS leaves the city. Divided into eight parts, Wojnarowicz's book reads at times like the memoir of its subtitle and at others like a manifesto, political essay, or impressionistic film. If *North Enough*'s leaps across time and place are more or less traceable, *Close to the Knives*'s are often too erratic to chart. With few markers fixing time, place, or person, the narration has an oneiric quality that undercuts certainty about whether the book strictly recounts Wojnarowicz's own past. The narrator states, "there is really no difference between memory and sight, fantasy and actual vision," and the memoir's narrative style accords with this principle.[56] For this reason, the book's status as nonfiction is ambiguous. (To underscore this uncertainty, I alternate between "Wojnarowicz" and "narrator" when referring to the source of the narration.) Grover's text expands the reach of memoir through a polyphony that includes the voice of nature, but it remains historically, biographically, and ecologically referential. *Close to the Knives* expands the genre even further by destabilizing referentiality and repurposing memoir as a weapon in the broiling social and political battles that *Close to the Knives* contours. In this sense, the book more dramatically bends a genre that Stacy Alaimo identifies as primarily "a site for private self-reflection," one suitable to tales of individuals and families.[57] With these formal characteristics, the memoir's texture is as rich as the multimedia art for which Wojnarowicz garnered fame before his death from AIDS complications in 1992.

Beginning in the late 1970s, the artist created controversial, sometimes erotic paintings and photographs to counter the mainstream media's and politicians' derogation of queer sexualities. Though Wojnarowicz was entrenched in the 1980s New York art scene, not all of his art was metrocentric. Two pieces in particular help us unglue the artist from his customary milieu. The first adorns the cover of *Close to the Knives*. *Untitled (Buffalos)* (1988–1989) suggests a community careering toward oblivion against the backdrop of a grand Western landscape (figure 2.1). A photo of a diorama in the Smithsonian Institute, it evokes the imperialist United States's long history of decimating its own riches as a precedent for the "disintegrations" occurring in the late twentieth century. The everyday lives of homosexuals, PWAs, artists, and anyone living outside the bounds of sanctioned normalcy are as precarious as that of the Plains buffalo. In *Delta Towels* (1983) Wojnarowicz samples the letterpress grocery circular, ubiquitous in mid-century small-town America, and splices in woodcuts representing bestiality (figure 2.2). Like the other works in his supermarket series, this piece's nostalgic visual vocabulary literally foregrounds the so-called perverse pleasures found around the corners and behind the barns of rural America. Created in the first years of the AIDS crisis, *Delta Towels* draws attention to the varieties of sexuality that a heteronormative nation rarely recognizes, much less sanctions.[58] The geographical reference in the title locates viewers in the South, and the choice of the black-and-white woodcut racializes the human figure in the piece. Like the stories that E. Patrick Johnson collects in *Sweet Tea: Black Gay Men of the South*, *Delta Towels* attests to the "queer possibilities in the South," particularly in nonurban spaces, and contests the presumption that these spaces are necessarily "inhospitable" to nonhetero desire and sex practices.[59]

Such nonurban imaginaries energize Wojnarowicz's disruptive work, and they slightly fade our picture of him as an exclusively city artist. Composed in the late 1980s and published one year before Wojnarowicz's death, *Close to the Knives* also situates the artist outside of the city. The memoir documents both the social meanings of HIV/AIDS when it was first ravaging the New York arts world and the author's personal history of violence, desire, loss, and self-exploration. Across eight sections that range from the exhilaration and disillusionment of Jack Kerouac's *On the Road* (1957) and William Burroughs's "cut-up" novels of the 1960s, to the vehemence of a political

FIGURE 2.1. David Wojnarowicz, *Untitled (Buffalos)*, 1988–1989. Gelatin silver print. 40½ x 48 inches. *Courtesy of the Estate of David Wojnarowicz and P.P.O.W. Gallery, New York*

tract, the book travels between city, suburb, and undeveloped spaces. Traversing these zones, it recounts dreams and "calculated fucks" (3), hospital visits and road trips, myths and statistics about AIDS. At full voice and with fury, Wojnarowicz condemns what he calls the "one-tribe nation" for its attempts to destroy artistic, sexual, and environmental difference in America (37).[60] His inflammatory rhetoric, which pointedly contrasts Grover's sometimes angry but always measured voice, is rooted in its historical moment. This was a moment of confusion and rage before highly active antiretroviral therapies became widely available after the Eleventh International Conference on AIDS in 1996 and made the disease more livable for Americans with means.[61] I read *Close to the Knives*'s environmental relations as key to how it expresses this rage and develops its politics of resistance. Wojnarowicz's symbolically rich environmental positioning is integral to how the book recodes nature in the context of pervasive

FIGURE 2.2. David Wojnarowicz, *Delta Towels*, 1983. Spray paint and stencil on supermarket poster. 50¼ x 35 inches. *Courtesy of the Estate of David Wojnarowicz and P.P.O.W. Gallery, New York*

sickness and draws the contours of social and spatial injustice in the U.S. Similar to *North Enough*, the meanings of the losses that Wojnarowicz chronicles only fully emerge when the sick body enters new surroundings, and this encounter elicits a discordant conflict between expectation and response.

As in Grover's and Verghese's memoirs, personal mobility is the condition of possibility for Wojnarowicz's environmental reflections, but, for the latter writer, mobility—specifically, automobility—also abets detachment. Roaring down the interstate and navigating city streets, Wojnarowicz uses the car to think through the ethics and politics of urban planning. The narrator observes, "All along the sidewalks were the people reduced to walking; the desperation of whole families sitting in lethargy on the curbsides lost to the sounds of automobiles" (31). He has some faith that, if their feet were on the ground rather than on the pedal, drivers would share space with the impoverished and would thereby empathize with their plight. But, "owning a vehicle, you could drive by and with the pressure of your foot on the accelerator and with your eyes on the road you could pass it quickly ... so that these images were in the past before you came upon them" (31). Inequity and suffering cannot penetrate windshields to reach the detached individuals behind them. Along with the housing subdivisions that it helped spawn, the automobile promotes middle-class indifference and is a metonym for it.

Though Wojnarowicz does not explicitly address PWAs here, the circumstances of the carless—constrained mobility, "lethargy," and enforced silence in the roar of fast-paced life—rhyme with those of the severely ill. Sickness positions the artist to perceive the exclusions that are built on economic and spatial disparities and that surface in the American landscape at century's end. Journeying along the highway, city sidewalk, or suburban street, the body that knows sickness knows these disparities and exclusions, and the significance of those spaces is altered. *Close to the Knives* thus makes a claim for the imbrication of body and environment in two complementary ways. It establishes that figuring sick bodies in literature necessarily changes figurations of space. Simply put, the eyes of the ill do not perceive the same land as the eyes of the healthy.[62] *Close to the Knives* argues that "setting" is not background but is a force eliciting affects like discord that make ways of seeing possible. This perceptual adjustment indicates a second way in which body and space interanimate: the memoir demonstrates that land development patterns and social responses to disease share an ethic that rests on discrimination and neglect.

As the narration moves to the suburbs, these two forms of body-environment involvement come to light. Though *Close to the Knives* begins in the city, Wojnarowicz's environmental biography begins in this milieu. The suburbs are the "Universe of the Neatly Clipped Lawn . . . where anything and everything can and does take place—and events such as torture, starvation, humiliation, physical and psychic violence can take place uncontested by others, as long as it [sic] doesn't stray across the boundaries and borders as formed by the deed-holder" (151). The violence Wojnarowicz describes is not abstract: elsewhere in the memoir he tells how suburban propriety helped shelter his own father's abuses. "Subdivision" here refers not only to the parceling out of space for developers' profit but also to the parceling out of compassion. Passing through unnamed suburbs on a road trip as an adult, he observes the reification of private property under the American Dream. Imaginary and physical boundaries, symbolized by "the neighbors [sic] split-rail fencing" (60), both hide the private violence taking place within their bounds and inure the sheltered residents to the injustices that take place outside them.[63] Wojnarowicz asks incredulously, "Do they believe that the virus stays within the boundaries of large urban centers even though this is a country of trains and planes and automobiles?" (158). The naive but also injurious belief that HIV/AIDS and other stigmatized diseases only affect "other" people is one of the injustices that the suburbs perpetuate.

What Lewis Mumford describes as the suburban "temptation to retreat from unpleasant realities, to shirk public duties, and to find the whole meaning of life in . . . the family . . . or even the self-centered individual" does not appeal even to the young Wojnarowicz.[64] Rather than retreat to the nuclear family, he takes shelter from violence in the alternative "universe" of the woods (152). He first administers the same "geographic cure" that allured Grover and Verghese. "Once I discovered the universe of the forests and lakes," he recalls, "I went there whenever possible. In the universe of the forest I didn't think about the Universe of the Neatly Clipped Lawn" (152).[65] Wojnarowicz approaches the woods not as a naturalist but as a sort of environmental pilgrim who yearns to resonate with planetary harmony. With this desire, he follows in Transcendentalists' footsteps and demonstrates the persistence of the idea that nature strikes universal chords. But he doesn't need the latest ecological theories to teach him that harmony is an antiquated ideal. He laments his belatedness and birth into a fallen world. As late twentieth-century Americans, we necessarily inhabit "a place where by virtue of having been born centuries late one is denied access to earth or space, choice or movement. The bought-up world; the owned world" (87). Even wilderness, the space through which Americans have imagined an alternative to civilization, is inaccessible. Rather than enumerate the materials of industrialization and technological modernization that obstruct this access, Wojnarowicz names conceptual obstacles. Nature *as concept* guards the gates to the harmonious, salubrious nature that we were taught to value. "The invention of the word 'nature' dissociates us from the ground we walk on," the narrator elaborates (88). "Nature" differentiates what Neil Smith terms "first nature," "a blind conceptless occurrence," from "second nature," "the world of men as it takes shape in the state, law, society, and the economy."[66] Petrified ideas of nature then are barriers to environmental connection and, in the memoir's logic, interpersonal connection.

In title, and in angry content, *Close to the Knives* is a far cry from the kind of nature writing that appears in Grover's memoir. Yet Wojnarowicz occasionally shares with nature writers a quest for connection to and even oneness with the nonhuman, a quest he pursues in more and less developed spaces. Returning to New York City's liminal zones—a pier, a warehouse, a diner—*Close to the Knives* narrates occasions when a person merges with human and more-than-human others in states of sensual transport. These zones have a fluid, erotic character that the present progressive, sensuous narration conveys:

> Cars bucking over cobblestones down the quiet side streets, trucks waiting at corners with swarthy drivers leaning back in the cool

> shadowy seats and the windows of buildings opening and closing
> . . . faraway sounds of voices and cries and horns roll up and funnel in like some secret earphone connecting me with the creaking movements of the living city. Old images race back and forth and I'm gathering a heat in the depths of my belly from them: flashes of a curve of arm, back, the lines of a neck glimpsed among the crowds in the train station. (12)

Beginning with the verb *bucking* and the suggestive posture of "swarthy drivers," the theme of connection takes on a sexual tone that the "secret earphone" amplifies.[67] Here, as elsewhere, the motif of merging with the environment appears in unlikely places. In the vignette that opens part 3, "IN THE SHADOW OF THE AMERICAN DREAM: Soon All This Will Be Picturesque Ruins," the narrator somatizes the landscape to express land as body. While ostensibly liberated by the road, Wojnarowicz equates himself with a prisoner caged in a "ruined" nation.

> In the watery circling of shapes and textures, I saw pieces of anatomy surfacing from my sleep: the lips or cheekbones or the fingers of some man or woman speaking and there was no sound but I recalled some story about a man lying in a prison cell. . . . A small window high up on the wall across from his bed allowed him on tiptoe a view of a tiny piece of landscape, the tip of a rock or the shallow hip of hillside. . . . One day he discovered that he could measure the distances of the landscape by lying on his back in the center of the floor and placing the soles of his bare feet against the shafts of sunlight extending diagonally through the bars. (25)

Prison is the unlikely setting for a vision of positive embodiment. It is possible to be in the body *through* landscape, to envision a way out of the embattled embodiment that homophobic and anti-AIDS rhetoric prescribes. Anatomical metaphors such as the hill's "hip" and metonyms such as the prisoner's topographical survey via the body set up the promise—if not the fulfillment—of unmediated connection to the land and foreshadow *Close to the Knives*'s evanescent moments of merging.[68]

The memoir evokes the merging narrative prevalent in environmental writing from John Muir to Leslie Marmon Silko's *Almanac of the Dead* (1991), the focus of chapter 5. Silko's novel innovates with the merging trope by routing it through scenes of violence; Wojnarowicz's text updates it by proposing that connection is ephemeral because of the conceptual tyranny of "nature." In the U.S., "nature" is owned by a propertied class that codes it as pristine and healthy and uses it as both a tool of and shelter for oppression. When Wojnarowicz's sick, queer body enters the majestic landscapes of the American Southwest, the next stop in his environmental itinerary, this conceptual tyranny hits him even harder and provokes a bitter, irritated response. In contrast to Grover's finely detailed northern Minnesota, Wojnarowicz's New Mexico is only an outline. The narrator does not place readers on a map, catalog flora and fauna, or recount the economic and historical contingencies that have produced the space. In this respect, *Close to the Knives* partially levels geographical difference. However, like his fellow AIDS memoirist, Wojnarowicz's environmental encounter produces forceful but subtle irritations of discord that disrupt the epistemic and perceptual habits governing environmental experience.

If Falwell and others used the concept of nature to condemn the immorality of queer desire, *Close to the Knives* uses encounters with nature as a window onto the disintegration of "**A DISEASED SOCIETY**" (114). One such encounter takes place in the desert that the narrator enters in the memoir's sixth part, "POSTCARDS FROM AMERICA: X Rays from Hell." For Wojnarowicz, as for Silko, the desert is fertile ground for exploring aesthetic responses to environments seen through the frame of sickness. He takes in a range of mountains with a filmic gaze that contrasts with the static medium of sentimental nostalgia, the postcard. As his eyes sweep across the range, his fury mounts, and he rejects nature's aesthetic and salutary appeal.

> When I was out west this summer standing in the mountains of a small city in New Mexico I got a sudden and intense feeling of rage looking at those postcard-perfect slopes and clouds. For all I knew I was the only person for miles and all alone and I didn't trust that fucking mountain's serenity. I mean it was just bullshit. I couldn't buy the

> con of nature's beauty; all I could see was death. The rest of my life is being unwound and seen through a frame of death.... **WHEN I WAS TOLD THAT I'D CONTRACTED THIS VIRUS IT DIDN'T TAKE ME LONG TO REALIZE THAT I'D CONTRACTED A DISEASED SOCIETY AS WELL.** (113–14)

Though the diction and tone of this passage are inconceivable in *North Enough,* the narrator here echoes Grover in detecting beauty even within a death-tinted landscape. The "mountain's serenity" appears everlasting, and this grates on Wojnarowicz's sense of mortality. HIV is communicable in several senses, these writers show; it infects bodies through contact, but it also spreads into the spaces and social milieus those bodies inhabit. Indeed, in the memoir's concluding part, the narrator iterates his confession that he is "preoccupied with the sense of disease and death in the environment" (227). However, Wojnarowicz's sense of death in the environment cohabits with a more typical, affirmative reaction to the southwestern tableau. Though the mountain elicits anger, Wojnarowicz still perceives the mountain's beauty as "postcard-perfect," much as Grover initially adhered to the script of apprehending the devastated Minnesota forests and bogs as ugly. He does not marshal negative topoi of the desert as barren wasteland in order to think through death and disillusionment. Just as "neat" suburban property stands in for an ideology of isolationism, inequity, and neglect, the touristic "beauty" of the desert is part of a representational system that naturalizes the alignment of normalized aesthetic, environmental, and medical values. With the "more onerous citizenship" that sickness confers on him, he relinquishes rights to the other terms in the beauty-health-nature equation.[69] When environmental experience is filtered through the virus—through the material body or "frame of death," in Wojnarowicz's idiom—the sedimented values of nature diffuse.

If artists and critics have noted that the spaces of AIDS change how people live with sickness, Grover and Wojnarowicz add that sickness changes the meanings of space in turn. As I will elaborate further on in this chapter, this change takes place through the prism of affect. The artist-traveler feels discord between his momentary aesthetic appreciation of the mountain's beauty and his feeling that the conventions governing such beauty are part of a regime of thought about his environments that he otherwise denigrates. This discord reconfigures Wojnarowicz's conceptual map for nature. It is no longer a space of harmonious merging, a retreat, or an analogue for the healthy body, but, as an ideologically-charged concept, is instead an analogue for social and spatial injustice.

The discord that arises between Grover's and Wojnarowicz's initial reactions to their environments and their lived experiences of disease and space is a peculiarly environmental emotion that merits more attention. Discord opens up fissures in the entrenched semantics of "nature" and denaturalizes its equation with health and beauty. An affect elicited by sickness, discord confers on the memoirists perceptual and epistemic agency. It calls out normalized patterns of apprehending bodies and environments and provides the means for severing the conceptual chain that binds them around an idealized harmony. The memoirists come to distrust the "con" of preestablished environmental perception—of sanitized suburbs as well as picturesque wilds—and to cherish so-called waste places.

DISCORDANT FEELINGS, SUSPICIOUS STANCES

Reading interactions between sick bodies and nonurban landscapes in *Close to the Knives* and *North Enough* shows discord to be a peculiarly environmental affect. The expectation that nature is harmonious, healthy, and aesthetically pleasing informs Grover's and Wojnarowicz's environmental dispositions. Though the tropological regime surrounding "nature" persists as they leave the city, their sickness experiences activate the discord that eventually recalibrates the ways in which they apprehend earth and soma alike. The disturbances of one type of radical human-nonhuman contact—that between an individual and the human immunodeficiency virus—are the precondition for later discordant environmental experience. Arising from sickness, discord shows that the envelope of the body cannot be shed within ecological thought since environmental relation necessarily filters through the body. Writing the interface between sickness and nonurban environment, the memoirs announce that it is always an embodied person—in different states of health, in different class positions, with different types of sexual desire—who perceives his surroundings and sets his obligations toward them.

What are the workings of discord such that the emotion productively transforms one's epistemic, aesthetic, and ethical bent toward the world? *Close to the Knives* and *North Enough* corroborate Charles Altieri's assertion that "many emotions either constitute values or call our attention to values because they position us in distinctive ways toward what unfolds in immediate situations."[70] In Grover's and Wojnarowicz's memoirs, discord announces a problem with the values that underpin but also overwhelm their dispositions toward nature. But while they support Altieri on this point, these works also modify his model. Whereas he emphasizes that affect draws attention to what unfolds *in* situations, they also show that affect draws attention *to* our situation in space. Sickness positions the memoirists in "distinctive ways" toward their surroundings, and these positions generate the discord between established and emergent perceptions and evaluations of environment, body, and beauty. With this effect, discord has an intense quality that "leads us to posit our present [affective] state as set over against ordinary epistemic practices. Intensity has to be measured in terms of the degree to which it allows us to engage the world differently from the practices we adapt when we are operating on the same practical plane as those around us."[71] While not one of Philip Fisher's "vehement passions" like rage or terror, discord is intense because it disrupts customary ways of seeing and thinking.[72] In effect, the affects that most rock us are most likely to rock the foundations for evaluation. By this measure, discord has an intensity belied by its subtlety. It exposes the extent to which "ordinary epistemic practices" are naturalized and internalized; it lays bare the aesthetic and conceptual templates that we don't always know we're bringing to bear on unfolding encounters. It is an affect particularly apt to pinpoint the troubling persistence of values that, once questioned, must be reconfigured. Discord is epistemically unsettling, then, because it impels us to scrutinize automatic filters for knowing, assessing, and living with the world around us.

Discord unearths the fault lines between expectation and experience because it bridges gut and mind. Grover and Wojnarowicz do not, as I am here, study or think their way out of the petrified tropological chain that hooks nature up to health and beauty; instead, they *feel* a disturbance. As persons touched by disease enter suburb, wilderness, bog, and forest through the frame of sickness, the conventional meanings of nature assert themselves. They initially respond to their new surroundings according to existing metrics of value: marshes and mountains are homely or beautiful, used up or perfect. Reacting thus to new landscapes, the memoirists discover their own complicity in perpetuating epistemic and evaluative positions that they themselves vilify. The immediacy of their responses to the north woods and the U.S. Southwest demonstrates that they have internalized the conceptual edifice that constitutes "nature." However, in a subsequent moment, discord jars these initial habits and works as affective resistance to ingrained patterns of thought.

On these points, the comparison to *My Own Country* stands out. Even though Verghese's memoir similarly expands the geography of AIDS, "nature" has a conceptual and affective hold on him. Projecting nostalgia and harmony onto rural Tennessee, Verghese sees his environment as an avenue to belonging. He imagines himself fitting with a country whose components are themselves in harmony, and his first perceptions of rolling hills, pastureland, and native oak trees endure. By contrast, Grover's and Wojnarowicz's feelings of discord dig up and help dismantle the conceptual foundations on which their evaluations of the more-than-human world are built. They reject their primary responses because those responses conform to a conceptual regime that is at the root of the environmental, biomedical, and social injustices that they oppose ethically. Discord does not only destabilize these AIDS writers, however; it also releases an epistemic agency previously unavailable to them. The structure of *Close to the Knives* suggests that, after the narrator recognizes nature's "con," he is ready to unleash the targeted vitriol that fills the two subsequent sections of "POSTCARDS FROM AMERICA." Wojnarowicz writes with the hope that "to speak about the once unspeakable can make the **invisible** familiar if repeated often enough in clear and loud tones" (153). While the first sections of the memoir cleave to personal and sexual anecdote, the sections after "X-Rays from Hell" disrupt consensus knowledge in "clear and loud tones" that attest to Wojnarowicz's own awareness of the formerly "invisible" commitments that discord has made visible.

The intensification of Wojnarowicz's polemic is a rhetorical upshot of discord, but there is also a formal correlate to the emotion that unites the two writers' texts: polyphony. As mentioned previously, *North Enough* and *Close to the Knives* adhere to readers' expectations for the memoir genre in many respects, but violate those expectations by multiplying voices. On the one hand, these works are primarily narrated in the first person and validate the authority of embodied experience. Yet Grover and Wojnarowicz also ventriloquize other voices as they put themselves in dialogue with sociopolitical, human biological, and environmental forces. Open to other voices, these texts stage their arguments about human vulnerability to disruptive invasions of the more-than-human, disease included. Like the "material memoir" that Alaimo details, the discordant AIDS memoir "undertakes [a] sort of self-questioning" and "reveal[s] how profoundly the sense of self is transformed by the recognition that the very substance of the self is interconnected with vast biological, economic, and industrial systems."[73] Thus, even though the memoirs promote embodied ways of knowing the land use trends that shape the earth and themselves, their narratorial openness destabilizes the self around which memoir is usually organized.[74]

In effect, discord sets us outside of ourselves such that we become checks on our own habits of thought. The self-questioning stance of these memoirs thus amounts to a suspicion that is crucial to the implications of discord for thought and ethics. To put a finer point on it, we can separate out discord's affective and epistemic dimensions. If discord, a disturbance between immediate response and experience-shaped evaluation, is the affective content of irritation, suspicion is discord's epistemic content. In effect, suspicion is essential to the move from ecstatic discord to new knowledge positions. After we recognize deep-seated patterns of thought, discord erupts to produce a suspicious stance from which to assess those disquieting normalized narratives of, in this case, nature. Suspicion permeates Grover's and Wojnarowicz's texts. It resounds in Wojnarowicz's denunciation of the mountain's beauty as a deception. Earlier in *Close to the Knives*, suspicion intensifies precisely in the moment when the narrator has a fleeting hint of the environmental harmony that the forests seemed to promise. Attuned to "the movements of the planet in its canyons and arroyos, in its suburbs and cities," he detects in the whirl "a spark so subtle and beautiful that to trust it is to trust our own stupidity" (69). Across both memoirs, if a judgment comes to hand too quickly, it signals a failure to evaluate the norms and conventions that inform that judgment. In Grover suspicion also arrives through aesthetic reflection; she acknowledges she has "learned to be deeply suspicious of metaphor, resistant to the pretty conceits that once satisfied my need to explain pain. When I look south to the surviving pines, I try to abjure the lessons that spin so readily to mind" (21). Grover's comment indicates the temporal dimension of suspicion: it slows down processes of perceiving, evaluating, and representing the world around us, even to ourselves. This *ritard* is crucial to the discord-suspicion mechanism because, in that extended temporality, the affect produces an epistemic ecstasy born of embodied environmental encounter.

In the model of discord that these AIDS memoirs propose, suspicion escorts new understanding because it puts the brakes on thought and encourages epistemological self-examination. The sanguine picture of suspicion that I draw here sharply contrasts the critical debate on this disposition. I put nonurban AIDS memoirs in conversation with cultural critics who have become suspicious of suspicion in order to clarify the discord mechanism and show the merits of suspicion as a worldly stance. Eve Kosofsky Sedgwick, Bruno Latour, Rita Felski, and Anne Cheng have urged fellow scholars to reconsider suspicious reading because it constrains interpretive and critical practices and impedes ethical ways of being.[75] Paul Ricoeur's "hermeneutics of suspicion," outlined in *Freud and Philosophy* (1970), first articulated "a new relation . . . between the patent and latent," an interpretive project "to look upon the whole of consciousness as 'false' consciousness."[76] Sedgwick opposes Ricoeur's paradigm, which requires that we interpret a text according to what is hidden beneath it, out of concern for what this hermeneutics brings about. While "very productive," she admits, the hermeneutics of suspicion "may have had an unintentionally stultifying side effect: they may have made it less rather than more possible to unpack the local, contingent relations between any given piece of knowledge and its narrative/epistemological entailments for the seeker, knower, or teller."[77] Though rejoinders to suspicion such as Sedgwick's target the ways we do criticism, criticism is often a metonym for epistemic and ethical orientations that manifest beyond the page. This is evident in the "/" that equates "narrative" and "epistemological entailments" in Sedgwick's sentence above. Felski also collocates text and episteme, interpretation and ethics, when she explains that "modes of critical thought are also forms of orientation toward the world, shaped by sensibility, attitude, and affective style."[78] For Sedgwick, knowledge born of suspicion is ultimately not felicitous; it does not alter one's bent toward the world or lead to material changes in it because of its strong "side effect": paranoia.[79] An alternative to the "stultifying" paranoid reading practice that suspicion begets is a generous receptivity to whatever may come. Cheng dubs this a "hermeneutics of susceptibility, rather than suspicion. By this I mean a reading practice that is willing to *follow*, rather than suppress, the wayward life of the subject and object in dynamic interface."[80]

In Felski's compatible description of this antisuspicious stance, it involves "delving into the mysteries of our many-sided attachments to texts" which include "trance-like states of immersion or absorption . . . surges of sympathy or mistrust, affinity or alienation . . . sensations of fretfulness, irritation, or boredom."[81] Felski gives the most detailed reckoning of suspicion as itself an affect with epistemic implications. In her account it is both a cognitive state with an affective signature (that is, "states" in the quote above) and the affect itself (that is, "sensations"). After defining suspicion as "not just a cognitive exercise but an orientation infused with a mélange of affective and characterological components," she redescribes it as "a curiously non-emotional emotion, a quasi-invisible affective state that overlaps with, and builds upon, the stance of detachment."[82] In other words, suspicion is a disposition that arises from a cocktail of eliciting emotions and personality traits, but it is also an affect in its own right. My analysis of AIDS memoirs upholds the idea that suspicion is an epistemic orientation, but it asks us to reconsider this claim that suspicion *is* the affect. *Close to the Knives* and *North Enough* instead show that suspicion is an epistemic by-product of discord. The disjunction felt between initial perceptions and subsequent reflection on the preexisting norms that shape those perceptions ushers in a suspicious mode of thought.

Once we grasp this distinction between affect and epistemic disposition, suspicion can have interpretive outcomes other than paranoia and disillusionment. When suspicion *precedes* experience, it might very well lead to the limiting, one-sided interpretations that trouble Sedgwick and others. However, when it arises out of the decidedly embodied emotion of discord that Grover's and Wojnarowicz's texts theorize, suspicion can yield precisely those dispositions that the critics endorse: ones marked by shock, susceptibility, and enchantment that move us to fashion "distinctive configurations of social *knowledge*."[83] For Grover and Wojnarowicz, the reconfiguration of knowledge emerges from the discord-induced suspicion they experience when sickness frames their engagements with their environments. In the nonurban AIDS memoir, suspicion ushers in surprise rather than impeding it. Northern plains, landfills, suburbs, and mountains all become other to themselves through felicitous discord. A codified logic governing nature endures, but discord makes available "local, contingent" meanings of body and environment, sites previously understood through rigid conceptual equations.[84] The more-than-human world accrues value through a calculus of loss and compensation, and a recoded "nature" encompasses ugliness, dying, pain, and disorder. The revised understandings of nature, beauty, and health that arise with discord and suspicion disclose rather than foreclose possibilities for interpretation. The suspicious stance in the AIDS memoir of discord is not an obstacle to seeing the world anew; it is a pathway to reorienting environmental and somatic knowledge and ethics.

In this respect, discord is a type of "unrest that makes knowledge move."[85] Perceptual and epistemic practices change as somatic sickness enters nonurban environments and discord uncovers the concepts and habits of thought that have determined the body's and the earth's value to that point. To put the discord mechanism schematically: seeing irritates expectations for the "ideal" appearance and function of bodies and ecosystems, and this discordant feeling in turn positively irritates understanding of these domains. Discord sets us on the path toward what Timothy Morton terms "ecology without nature," that is "'ecology without a *concept* of the natural.'"[86] If, as Morton contends, ideological thinking "tends to fixate on concepts rather than doing what is 'natural' to thought," then the environmental affect of discord can unfix those concepts by derailing normalized thought.[87] *North Enough* and *Close to the Knives* do not set in stone a new conceptual pattern or insist on universal consensus, but there is still an important political dimension to the discord mechanism these texts model. Discord does not only serve to redraw a person's conceptual map of "nature;" it also validates firsthand experience as an instrument of knowledge and ethics. As the "narrow social frame" that usually encloses memoir widens, the sick body as environmental interface accrues legitimacy and undoes tropes of nature that circulate in popular as well as environmental and medical discourse.[88]

DISCORD IN ACTIVISM

Wojnarowicz's work announces its political commitments most vociferously. "Each painting, film, sculpture or page of writing I make represents to me a particular moment in the history of my body on this planet, in america [*sic*]," and telling this embodied history "open[s] up certain boundaries and releases information that unties the psychic ropes that bind the ONE-TRIBE NATION" (149, 143). As I have argued, discord is key to Wojnarowicz's own experience of the kinds of "psychic" disturbance that he hopes to effect in his audience. It is also key to two contemporaneous activist movements that complement Wojnarowicz's and Grover's artistic explorations of discord. To conclude, I address the AIDS health and environmental justice movements that share common cause with *Close to the Knives* and *North Enough*.[89] Like the memoirists' projects, these movements lend credibility to local, experiential knowledge as a form of authority that is not imperious. Discord and suspicion do not simply take down an established regime of thought; they also install bodies as material and subjective grounds for epistemic legitimacy.[90] When we read Wojnarowicz's and Grover's memoirs alongside activist projects that use sickness experience to question dominant medical and environmental paradigms, we better see the political significance and historical specificity of their writing.

In the midst of the AIDS crisis, PWAs questioned biomedical paradigms for determining whether someone is ill, the sources of the illness, and how the sick should be treated. This questioning arose from a felt discord between life with AIDS and how the media and institutions of public health represented and sought to manage that life. In particular, AIDS activists held the medical establishment under suspicion because of the slow pace at which public health policies, research agendas, and protocols for care emerged in the late 1980s and early 1990s and because of the prejudice and hostility toward those populations exhibiting disproportionate prevalence of the condition. The AIDS health movement took cues from the feminist health movement of the 1970s, which, according to medical sociologist Steven Epstein, also sought to "advocate skepticism toward medical claims" while remaining "'pro-knowledge'" rather than "'anti-science.'"[91] In order to balance respect for science and skepticism toward it, activists had to balance trust and suspicion. For example, most PWAs and activists "accept that a solution to the deadly AIDS epidemic will arrive via some variety of scientific process, if it arrives at all. They are caught, as Ronald Brayer puts it, 'between the specter and the promise of medicine.'"[82] Discord is thus central to the medical politics of AIDS, as it appears not only where experience and research data conflict but also where necessary trust in experts rubs up against knowledge of past and persistent malfeasance.[93] The AIDS health movement strikes another delicate balance: it advocates for health justice while allowing for multiple definitions of health based on embodied expertise. Even as they might endorse Article 12 of the United Nations International Covenant on Economic, Social, and Cultural Rights, which enshrines "the right to the enjoyment of the highest attainable standard of physical and mental health,"[94] they would also support Nietzsche's axiom that "there are innumerable healths of the body" rather than a singular "*normal* health."[95] Epstein relates that PWAs' confrontations with public health agencies, pharmaceutical companies, and medical researchers helped mint the body as one of the "new 'currencies' for credibility . . . within the scientific field."[96] Appeals to the (sick) flesh put force behind suspicion about expert knowledge that did not align with the lived experience of AIDS.

The 1980s also saw poor and racially marginalized individuals with ailments caused by industrial pollutants press similar lines of questioning based on their discordant response to consensus environmental and medical science. Like PWAs, those advocating for environmental health justice "have something to say simply because of where they stand."[97] Giovanna Di Chiro elaborates that local expertise depends on a person's keen attention to both her sick body and her environmental conditions. This twofold knowledge accounts for the success of much, though of course not all, grassroots environmental justice work in the U.S. and around the globe. Actual sick bodies empower challenges to prevailing wisdom about somatic and environmental health, and these challenges "call into question issues of epistemological legitimacy and, furthermore, may generate newly configured sites of environmental contestation."[98]

Who has the right to speak for the health of bodies and the earth? What codified concepts of the human body and the more-than-human govern understanding of health itself? At century's end, AIDS and environmental health activism give urgency to these questions, which Grover and Wojnarowicz approach in their life writing. Opening up fissures in petrified conceptual and representational systems governing "nature," *North Enough* and *Close to the Knives* contribute to the efforts of these "queer and environmental activists [who] have long . . . insisted that the redrawing of conceptual boundaries is intimately linked to the transformation of material practices involving both human and more-than-human natures."[99] The memoirists legitimate the epistemic, aesthetic, and ethical adjustments that result from the discord of living with sickness in varied environments. Yet they do not so much dictate new regimes of value as illustrate personalized, contingent ways of organizing understanding of earth and soma and obligations toward them.

Discord's value becomes clear under the pressures of ecosickness, of managing disease in socially normalized but in fact inharmonious spaces. Grover and Wojnarowicz share with health activists firsthand experience of a disconnect between lived sickness and the ideas of nature, health, and beauty that influence orientations toward environmental and biomedical dilemmas. Their memoirs of discord model a self-questioning that ushers in epistemological openness. They invite precisely the surprise, destabilization, and even questing for joy that form the nucleus of Sedgwick's reparative reading, Cheng's hermeneutic of susceptibility, and Felski's constructive enchantment. In nonurban AIDS memoirs, multiple voices attest that uncertainty, pain, and dying as well as surprise, appreciation, and adaptation are the very stuff of discordant (somatic and planetary) life.

Admittedly, discord does not always invite an easy ethics or politics. As I remarked earlier, Grover's acceptance of the as-is, an acceptance that derives from her reconceptualization of "nature," can hinder action against persistent logging, bog draining, and other practices that radically alter ecosystems. While acknowledging this difficulty, we can usefully reframe it through the approaches that the environmental justice and AIDS health movements have fashioned. These activists do not strive for absolute certainty and uncontested authority. After all, one-size-fits-all solutions would belie their commitment to the contingency of environmental and medical understanding. The goal rather is to cull multiple evidentiary sources that generate new questions for inquiry and strategies for representing affectively rich encounters between the human and the more-than-human. We can follow Craig Womack in branding this practice a "suspicioning" that explores "the relative merits of changing one's mind"[100] and welcomes the affective irritations—in this case, discord—that escort epistemic, ethical, and aesthetic upheaval.

North Enough and *Close to the Knives* lay the groundwork for my analysis of other ecosickness fictions in the coming chapters. First, these nonurban AIDS memoirs show that we must pay attention to the endurance of nature as concept, as "that which we are not" and as that which we essentially are.[101] This is not so much "the end of nature" as the end of stable conventions for seeing, representing, and caring for it. Second, they attest that, at the turn of the millennium, new narrative strategies such as memoiristic self-questioning take shape around affects that mediate individual, embodied experience and large-scale phenomena like global pandemics and land use. The affective disturbances that Grover and Wojnarowicz detail continue in Richard Powers's novels, to which I now turn. Whereas the AIDS memoirs promote the irritable affect of discord, *The Echo Maker* (2006) puts enthusiasm for the environmentally powerful emotion of wonder to the test. Nonurban AIDS memoirs violate readers' expectations about the memoir genre and about the epistemic, conceptual, and ethico-political work that discord can perform. Powers's novels also disturb expectations for early twenty-first-century environmental writing, but by narrating how an uplifting affect like wonder reroutes environmental and social care.

3

Richard Powers's Strange Wonder

In a 1672 letter to the Secretary of the Royal Society of London, Isaac Newton reports his observations of how a prism transforms light:

> It was at first a pleasing divertissement to view the vivid and intense colours produced thereby; but after a while applying myself to consider them more circumspectly, I became surprised to see them in an oblong form; which . . . I expected should have been circular. . . . Comparing the length of this colored spectrum with its breadth, I found it about five times greater, a disproportion so extravagant that it excited me to a more than ordinary curiosity of examining from whence it might proceed.[1]

A superficial pleasure, the distractive awakening of the visual sense, yields to play between the expected and the surprising, an "extravagance" that drives curiosity and keener investigation of refraction. Lorraine Daston and Katharine Park feature this anecdote in their study of wonder from the Middle Ages to the Age of Reason to emphasize that, without wonder, planetary mysteries would have remained just that, mysteries. It has long been the vital spark of inquiry into the matter of the world. How does a crystal break up a seemingly indivisible band of light? How can all of human existence rest on the proper functioning of microscopic cells, enzymes, and proteins?

How do migratory birds know the path to staging grounds they have never visited? The titans of the Scientific Revolution credit wonder with motivating study of unknowns such as these and lay the ground for building empirical research on a foundation of "a poetic sense of wonder."[2] In the eighteenth and nineteenth centuries, natural historians such as Carl Linnaeus, Gilbert White, and Henry David Thoreau put scientific wonder in the service of environmental awareness. As natural history develops into creative nature writing, wonder remains the hinge between individual perception and feeling and scientific inquiry. For the environmental thinkers who take inspiration from Thoreau and company, wonder is also what converts inquiry into care for our astonishing and increasingly threatened surroundings.

Given that wonder is a touchstone for natural history and nature writing of earlier generations, it's not surprising that it is still a guiding emotion for the twenty-first-century environmental imagination. What is surprising is that wonder has not yet received the same scrutiny as other affective commonplaces such as pastoralism and sentimentality. These dispositions have lost some of their shine in environmental discourse due to cultural critiques that reveal how pastoral idylls carry the taint of classism and elide the material relations between people and place.[3] Wonder, by contrast, has retained its status as an impetus to affirmative environmental relation. Reception of James Cameron's 2009 film, *Avatar*, attests as much. In one rapturous review, evolutionary biologist Carol Kaesuk Yoon pronounces the movie a "biologist's dream" because it sparks "the naked, heart-stopping wonder of really seeing the living world."[4] Specifically, it activates that play between recognition and surprise that also energized Newton. "With each glance," raves Yoon, "we are reminded of organisms we already know, while marveling over the new and trying quickly to put this novelty into some kind of sensible place in the mind."[5] Wonder overcomes the scientist and produces in her what René Descartes also identified in this "primitive passion": "a sudden surprise of the soul which makes it tend to consider attentively those objects which seem to it rare and extraordinary."[6]

In the twenty-first century, wonder endures, but its task is large. Wonder must not only shake apathy toward the more-than-human world and move us to curiosity without false idealization; it must also promote concern to curb the destruction of wildlife, of undeveloped space, and of human health and livelihood. The statistics on species loss show the scale of the challenge: as of 2012, 13 percent of bird species worldwide face extinction. Approximately 25 percent of land- and sea-based mammals are in similar distress.[7] The prognostications of Rachel Carson and fellow biologist Paul Ehrlich seem to be materializing. Birdsong is going silent. And, as the latter scientist wrote in 1968, "in spite of all the efforts of conservationists, all the propaganda, all the eloquent writing, all the beautiful pictures, the conservation battle is presently being lost."[8] In the war to prevent environmental collapse, Ehrlich worries that the weapons of marketing and art have not been powerful enough. Yet, forty years after his pessimistic assessment, environmental writers still pen wondrous accounts to dispel apathy. Richard Powers's ninth novel, *The Echo Maker* (2006), assumes this challenge, exploring whether, more than beauty and eloquence alone, unraveling the dynamic interactions between an individual's mind and the world it perceives might mold environmental consciousness.

Critics typically locate Powers within a postmodernist lineage, and for good reason. He violates the conventions of realist narrative through metafiction, builds many of his novels around poststructuralist reflections on linguistic meaning, and depicts how technology modifies perception and lived experience. *The Echo Maker* shares this postmodernist bent, as Rachel Greenwald Smith also argues, but these features serve to complicate motifs and principles that are integral to environmentalist thought.[9] Whereas his previous work *Gain* (1998) shares *The Echo Maker*'s interest in how human behaviors like using chemical household products compromise the health of unwitting consumers, the 2006 novel more directly addresses threats to the more-than-human realm. This chapter looks through the lens of wonder to see Powers's environmental commitments anew. This third novel, *The Gold Bug Variations* (1991) posits wonder about the building blocks of life itself as a counterweight to human hubris. It exalts wonder as that which redeems scientific enterprise when instrumentalism and the crushing proliferation of data threaten unadulterated curiosity. In *The Echo Maker*, wonder is the affective marker of the dual plots about ecosystemic and mental sickness. But even as the later text deploys and generates wonder, it delves more deeply into the affective mechanism as one that positions humans toward the more-than-human. Wonder brings to light how vulnerable bodies are metaphorically and materially imbricated in their at-risk surroundings. Yet, as the narrative details neurological function and the instability of consciousness, wonder is no longer understood as the salvific affect that it was in *The Gold Bug Variations* (hereafter, abbreviated as *Gold Bug*). Whereas *Gold Bug* asks, how can care arise from research into the shared genetic foundations of all life, *The Echo Maker* asks, how can we care for that which is other to us? Using and dismantling conventional understandings of wonder, *The Echo Maker* carves out an unorthodox place for itself within the tradition of environmental writing and nuances concepts such as place, connection, awareness, and care that animate environmentalist discourse. Revising these concepts, *The Echo Maker* probes how the mind mediates one's relation to the environment.

This chapter begins by laying out wonder's import to *Gold Bug* and then turns to *The Echo Maker*'s more vexed understanding of the affect. Like nature writing, the later novel aspires to incite readers' attention to their surroundings as a way to promote ecological protection. In line with this genre, it proposes that the affect of wonder seeds environmental projects. Wonder replaces causality as the hinge between the novel's two plots of sickness, one about species decimation and another about neurological injury. However, rather than simply urging one to wonder as *Gold Bug* does, *The Echo Maker* performs the affect. Elucidating ecological and neurological processes, the narrative bounces between lyricism and didacticism in an effort to defamiliarize everyday experience of self and place. As Newton's animated report attests, the play between familiarity and strangeness ignites wondrous experience; I demonstrate that this play is also the mechanism of cognition itself within the novel. Capgras syndrome, the disorder at the center of the text, proves this in the negative: it disrupts the mind's balance of the familiar and the strange. Thus, the structure of perceptual and cognitive processes and the structure of wonder are homologous. As *The Echo Maker* analogizes brain dys/function and environmental experience through the familiar-strange dialectic, it posits that systems operate thanks to interconnection, but it also undermines the ecological principle that "everything is connected." Ultimately, excessive connection making can lead to wonder's ugly obverses: projection and paranoia. Reimagining the consequences of connection in this way, the novel also inquires whether connectedness always entails care for human and nonhuman others. Complex affective arrangements thus derail ethical energies and disturb the care that wondrous awakenings promote in *Gold Bug*.

Powers's twenty-first-century ecosickness narrative deviates from the lineage of environmental writers that trust in the fidelity of the senses. It puts the baseline reliability of perception—and, thereby, awareness itself—under scrutiny. *The Echo Maker* is a rejoinder to platitudes about connectedness, but one that asks, can ethical concern take hold without connection and under conditions of what Dorothy Hale terms "psychological upset"?[10] A National Book Award-winning novel that appears on book group lists and university syllabi, *The Echo Maker* is vital for the critical project of *Ecosickness in Contemporary U.S. Fiction*: to determine the ways in which narrative affects establish the interdependence of vulnerable earth and soma and shape environmental investment. As Jan Zita Grover's and David Wojnarowicz's AIDS memoirs prove, a putatively negative affect such as discord can break apart conceptual chains and open new ways of valuing endangered bodies and spaces. Powers's works show wonder's potential not only to take us on this journey toward care but also to take us in the opposite direction. As the connections that wonder generates multiply and tip one into paranoia and projection in *The Echo Maker*, interconnectedness can in fact cut off generative relations of care.

"WEIRDLY ALIVE" WITH WONDER

With its focus split between the foundations of life and the foundations of music, *Gold Bug* continues a pattern to which all Powers novels conform: the interlacing of multiple plots that synthesize and humanize diverse domains of knowledge. Probing brain dysfunction and the ecological consequences of land use, *The Echo Maker* also fits this pattern and actualizes the question of narrative connectedness in plots *about* connectedness. The 1991 novel traces molecular biologist Stuart Ressler's fervent attempt to crack the genetic code with a team of researchers in the late 1950s and his eventual disaffection with laboratory science. His vocation ends when his lover leaves him to salvage her marriage and when utility and profit trump wonder as a biotechnological future looms. With sorrow, he anticipates that, "in a very few years, the Sunday-school work of cryptography will go public, enter commercial politics. Too much need always hinges on knowledge for it ever to remain uncorrupt, objective, a source of meditative awe. After wonder always comes the scramble, the applications for patents."[11] Leaving research behind, he finds a refuge in music, in particular Johann Sebastian Bach's *Goldberg Variations* which, along with Edgar Allan Poe's "The Gold-Bug," is an intertext for the novel's title. Ressler surfaces in the diegetic present of the novel, the early 1980s, when Franklin Todd approaches librarian Jan O'Deigh to help him uncover the mystery of the scientist's past. Now working with Todd to program and maintain banks of computers at Manhattan On-Line, Ressler shares his history as Todd and Jan draw him out of his reclusion.

From *Gold Bug* to *The Echo Maker*, readers follow the trends in scientific fascination: from genetic code in the 1950s and computer code in the 1980s to the brain in the 2000s. The 2006 novel sets an investigation of consciousness as the latest scientific frontier in an erstwhile frontier territory, Nebraska. New York neurologist Gerald Weber charts the former terrain. Lauded for his "neurological novelistic books," Weber narrates inconceivable brain disorders that reveal the precariousness of consciousness.[12] His works earn him both critical praise and censure that brings him to the attention of Karin Schluter, who is summoned back to her hometown of Kearney, Nebraska on 20 February 2002 when her twenty-seven-year-old brother, Mark, is injured in a car accident. After Mark emerges from his coma and regains speech, it becomes clear that his physical health belies a profound alteration to his mind. He has a delusional misidentification disorder known as Capgras syndrome: although Mark's memory remains intact, he no longer recognizes as real the people and objects that he loved most—Karin, his dog, and his modular home. He deems them impostors because he no longer feels for them what he once had. The usual affective attachments have been severed even as the perceptual pathways are unharmed. Though Weber has vowed to stop researching injured brains in favor of writing "an account of the brain in full flower," he cannot resist the lure of seeing "up close, through the rarest imaginable lens, just how treacherous the logic of consciousness was" (102). The tale of Mark's rehabilitation is set against two other events: the spectacular perennial migration of sandhill cranes through Kearney, and the battle between conservationists and developers for control of the Buffalo County Crane Refuge. Through the story of Kearney's land use contests, the text introduces Daniel Riegel, Karin's boyfriend and a devoted environmentalist who heads the Refuge, and Robert Karsh, Karin's lover and a developer with ambitions to turn the preserve into a water park. As *The Echo Maker* proceeds, narrative focus shifts from Mark's treatment to the environmental plot, a plot that eventually unlocks the mystery of why he drove his adored truck off a straight road on a cloudless night.

Across *The Echo Maker*'s modes of regional nature writing, detective fiction, domestic drama, and "neurological realism," we can isolate a primary objective: through a story about altered perception, the text seeks to alter readers' understanding of both human consciousness and environmental processes.[13] With this objective, it owes a debt to environmental writing that, reaching back to Thoreau and John Muir, has similarly cultivated awareness of local ecosystems by kindling wonder. In the twentieth century, Rachel Carson, Edward Abbey, and Mitchell Thomashow are just some of the writers who continue to posit wonder as crucial to environmental care. Wonder is a conduit to heightened perception and, as environmental educator Thomashow asserts, "is the basis for an ethic of care."[14] To combat the atrophy of perception and spread of apathy in the face of global environmental crisis, Thomashow pens a manual for "cultivating biospheric perception" and marshals wonder as the most powerful affect for encouraging planetary gratitude.[15] Carson's *The Sense of Wonder* (1965), a photographic and prose journal of her wanderings on the Maine coast, fits into Thomashow's curriculum. It chronicles her affective relation to her chosen home and exhorts readers to nurture "a sense of wonder so indestructible that it would last throughout life, as an unfailing antidote against the boredom and disenchantments of later years, the sterile preoccupation with things that are artificial, the alienation from the sources of our strength."[16] Just as Carson's sentence travels from vitality to alienation and despair, so too does the individual as she ages. Environmental experience, Carson proposes, can halt and even reverse the senescence of awe and liveliness. She anticipates Abbey's desert reflections on "the power of the odd and unexpected to startle the senses and surprise the mind out of their ruts of habit . . . into a reawakened awareness of the wonderful."[17] The canon of environmental writing is replete with thinkers who make cognate claims for wonder. Formally distinct from Thomashow's manual and Carson's and Abbey's life writing, *The Echo Maker* shares their project and develops their techniques for helping readers perceive their surroundings otherwise.

Simply put, Mark's injury, Capgras syndrome, is an extreme case of altered perception that ravages his life even as it opens his eyes to the radiant detail of the everyday. "Damage had somehow unblocked him," Karin remarks, "removing the mental categories that interfered with truly seeing. Assumption no longer smoothed out observation. Every glance now produced its own new landscape" (198). The fragmented perceptions of Mark's injured brain provide a template for how the text narrates his transformation. The result is a disjointed, clipped stream of thought that stages his recovery of speech: "Echo caca. Cocky locky. Caca lala. Living things, always talking. How you know you're living. Always with the *look*, with the *listen*, with the *see what I mean*. What can things mean, that they aren't already? Live things make such sounds, just to say what silence says better" (49–50). Speech indicates that Mark is recovering his humanity, but this alien speech discloses his distance from his former self and his fellow humans. His fragmentary, chiasmic thoughts sculpt their "own new landscape" (198). As the plot proceeds, neurological dysfunction stresses Mark's relationships and ruptures his sanity, but his splintered consciousness also activates new perceptions and flouts lay notions that sight and thought are smooth and whole. The question "what can things mean, that they aren't already?" imagines the possibility of speaking reality into greater significance. At the same time, his insights speak to a novelistic project of inducing a peculiar form of awareness of environments and their processes that shatters apathy.

Through the question of how to transform one's perceptions of the world, Mark's injury-induced abilities link up to the development battles in Kearney, which occupy part 4 of *The Echo Maker*. As Mark's only kin, Karin (whose orthographically distinct name evokes "caring") accepts responsibility for his rehabilitation, but sacrifices her identity when she abandons her life in Sioux City, Iowa. When Mark's condition worsens, Karin searches for something to salvage from futile caretaking and her diminished existence, anything "so long as it's uncompromised and wild" (407). She advances to head of public relations for the Refuge and, in her research, unearths a method for presenting the world anew, one that depends on facts ceding to bewildering sensation.

> She buries herself in legwork for the Refuge, researching her pamphlets. *Something to wake sleepwalkers and make the world strange again.* The least dose of life science, a few figures in a table, and she begins to see: people, desperate for solidity, must kill anything that exceeds them. Anything bigger or more linked, or, in its bleak enduring, a little more free. No one can bear how large the *outside* is, even as we decimate it. She has only to look, and the facts pour out. She reads, and still can't believe: twelve million or more species, less than a tenth of them counted. And half will snuff out in her lifetime. (407)

Raw data at first yields pessimism, and two difficulties erupt for Karin. She grieves that humans, always reactionary, will crush anything that exceeds them in magnitude, longevity, freedom, or biotic connectedness. The generic nouns *people* and *no one* become more personal when Karin focalizes the narration and includes herself in the collective *we*. Our reaction against the big produces the second problem, one of scale: how to comprehend the enormity of species extinction. While the scope of "the *outside*" may be unbearable, the vast scale of species and habitat decline incites an embodied response that might just curb the destructive impulse. The narrator continues,

> Crushed by data, her senses come weirdly alive. The air smells like lavender, and even the drab, late-winter browns feel more vivid than they have since sixteen. She's hungry all the time, and the futility of her work doubles her energy. Her connections race. She's like that case

> in Dr. Weber's last book, the woman with fronto-temporal dementia who suddenly started producing the most sumptuous paintings. A kind of compensation: when one brain part is overwhelmed, another takes over. (407)

The opening line of this excerpt achieves its full meaning when the passage concludes. When Karin can no longer process the information she is consuming, sensation springs to action and supersedes thinking. Time dilates across these lines. What appears to be a single snapshot of Karin reading enlarges when the narrator observes that Karin is "hungry all the time." Her senses expand as well: new aromas—lavender—and a new palette—"vivid" browns—enliven her surroundings and transport her back to youth. Finally, invoking the brain damaged painter, *The Echo Maker* proposes that aesthetic sense takes over where data fails. The aesthetic mediates the relation between a particular individual grounded in her sensing body and far-reaching environmental processes; it "*wake[s] sleepwalkers and make[s] the world strange again*."[18] So strange, in fact, that Karin renounces "everything human and personal" in favor of her environmental work (408).

Karin touches on a discovery that motivates many of Powers's novelistic projects: that scientists "can't get into all the implications [of their research] because the implications don't come out of well-formed questions and they're not all answerable by reductive, empirical programs. . . . There are places that empiricism simply can't get to."[19] Karin's epiphany and Powers's comments add to the scientists' reflections on wonder that opened this chapter. In Newton's account, for example, wonder initiates the desire to generate the kinds of facts that "pour out" of the tables Karin scrutinizes. *Gold Bug* concludes with Ressler's discovery that producing data testifying to human exceptionalism cannot be the directive of research: "The purpose of all science, like living, which amounts to the same thing, was not the accumulation of gnostic power, fixing of formulas for the names of God, stockpiling brutal efficiency, accomplishing the sadistic myth of progress. The purpose of science was to revive and cultivate a perpetual state of wonder."[20] Driven by wonder rather than hubris, the scientist will not pursue knowledge of life with the aim of "disenchanting the natural kingdom" and placing the human on its throne; she will work to establish that all creatures, like the genes that compose them, are "nothing in themselves but everything."[21] Wonder will arise from seeing the baseline commonality of all life and from finding in the randomness of humanity's slight distinction a reason to revere and care rather than to master. Crucially, *Gold Bug* announces "the purpose of science" through Jan, as she realizes that Ressler's musical compositions are a form of uncorrupted science. *The Echo Maker* shares *Gold Bug*'s concern that facts alone will ultimately destroy what is so wondrous. Karin "can't believe" the data and is startled by it, but the shock of the incomprehensible "crushes" her processing abilities. Rather than disabling her, this crush is liberating. Sensation rushes in, and takes her where data cannot. Thus, while wonder makes the ordinary strange, scientific research alone cannot sustain that defamiliarization. Sensuous understanding and aesthetic creation must take over.

This realization helps account for *The Echo Maker*'s narrative experiments with defamiliarization. Viktor Shklovsky's axioms about this modernist technique create a bridge between Karin's discovery about how cognition must cede to sensation and the form of the novel. In Shklovsky's Formalist account of defamiliarization (*ostranenie*, in the Russian), "art exists that one may recover the sensation of life; it exists to make one feel things, to make the stone *stony*. The purpose of art is to impart the sensation of things as they are perceived and not as they are known."[22] We can read Shklovsky's program as an aesthetic complement to Abbey's, Carson's, and Thomashow's perceptual projects in that the Formalist critic identifies what is linguistically necessary for activating surprise and shaking loose calcified response. Modernist writers follow the defamiliarizing program by flouting literary conventions, eradicating cliché from speech, and using sound and shocking imagery to create a new poetic language. Through such practices, Shklovsky asserts, literature infuses the ordinary with strangeness. "De-automatizing" perception and reaction, the literary work aims "to create a special perception of the object—*it creates a 'vision' of the object instead of serving as a means for knowing it.*"[23] Karin's revelations at the Refuge demonstrate Shklovsky's claim that perception supersedes knowledge in defamiliarization. Karin's epiphany induces an aesthetic urge, but does not derive from an artwork; it falls to *The Echo Maker* to defamiliarize Karin's ecosystem on its readers' behalf.[24]

On this point, it's important that Karin's ecopolitical commitments revolve around conservation and habitat restoration. The Refuge's mandate is to protect water resources for humans, but, rather than marshal water table statistics, it largely appeals to the spectacular beauty of the more-than-human—the cranes—and their somewhat inscrutable migratory and nesting behaviors. Karin enhances this strategy. Renouncing "everything human and personal" and redoubling her commitment to protecting her ecosystem (408), Karin arrives at a "'*vision*'" of habitat devastation that revolves around *impressions* of water's function rather than detailed hydrological *knowledge*. An image surfaces that crushes the "data" that threatened to crush her: "The web she glimpses is so intricate, so wide, that humans should long ago have shriveled up and died of shame. The only thing proper to want is what Mark wanted: to not be, to crawl down the deepest well and fossilize into a rock that only water can dissolve. Only water, as solvent against all toxic run-off, only water to dilute the poison of personality" (407–8). Water, repeated three times, reveals human insignificance as it erodes egoism. Water counteracts toxicity—an ability that "knowledge" of water would refute—as Karin imagines it to be an antidote to personality. Karin's experience supports the possibility that, with "the poison of personality" extracted, humans might awaken to establishing mutually salutary relations with the nonhuman. This scene shows aestheticized environmental wonder reviving sensation and fostering ecological engagement. Shklovsky's program thus balloons in *The Echo Maker*. Defamiliarization does not simply enliven perception of the everyday and draw attention to how art itself works. It also provokes reflection on the workings of the mind insofar as the account of Karin's revived sensation rhymes with Mark's awakening early in the novel (49–50). In other words, a "deranged" individual's mind and his curious perceptions provide a template for the revived awareness that Karin achieves. Modernist defamiliarization thus serves *The Echo Maker*'s arguably postmodern project of depicting "complex *processes* of reciprocity in which selves and environments come to bring about and shape each other."[25]

The narrative stylistics that generate wonder become clearer when we determine why wonder is essential to exciting perception and ethical involvement. One of Descartes's six fundamental passions and Aristotle's departure point for philosophy, wonder is notable for stimulating rather than stifling intellection. "When the first encounter with some object surprises us, and we judge it to be new, or very different from what we knew in the past or what we supposed it was going to be, this makes us wonder and be astonished at it. And since this can happen before we know in the least whether this object is suitable to us or not, it seems to me that Wonder is the first of all the passions."[26] Wonder, like the larger category of interest to which it belongs, "motivates exploration and learning, and guarantees the person's engagement in the environment."[27] We turn our eyes to fresh sights or see the ordinary anew in this affective state.[28] In the act of looking and marveling at an object or process—in Karin's case, sandhill cranes and their decimation and in Mark's, his own life refracted through Capgras—we become conscious of perception itself. We are astonished not only by the wondrous object but also by our capacity for awareness and our place in larger wholes. Directing thought to mental processes and the world that sparks them, this affect can channel investment in the object or event that gave rise to it. As Robert Fuller asserts, wonder orients an individual toward "sustained rapport with—and action on behalf of—the wider environment."[29] Because stories of ecosickness do not only imagine the imbrications of bodies in their local environments but also bridge the gap between small- and large-scale systems, it comes as no surprise that *The Echo Maker* experiments with wonder as care's helpmate.

The Echo Maker thematizes the production of wonder through Karin's conversion to conservation politics and formalizes wonder in a narrative mode that galvanizes the emotion. Diegetically and stylistically, then, wonder and practices of aesthetic defamiliarization are coupled. In particular, the text defamiliarizes everyday perception by joining storytelling, empirical thought, and poetic vision to spark readers' interest in neuro- and eco-biological systems. Specifically, it alternates between an expository mode, through which it unfolds elements of the plot and relates specialized understanding of brain function and habitat destruction, and a modernist, lyrical mode that, through unfocalized narration, adds mystery to technical knowledge.[30] The former mode adheres to what Powers calls a "closed, limited third-person focalization."[31] This is one of the respects in which the 2006 novel deviates from its 1991 precursor. Even as *Gold Bug* moves between Jan's first-person narrative set in the 1980s and Ressler's scientific *bildung* from the 1950s, the voice in these two sections remains consistent, as if Jan is, in effect, writing her friend's biography. Powers introduces more narrational variety into *The Echo Maker*. Elliptical passages filled with fragmentary, repetitive sentences open all but the third of the novel's five parts. In these preludes to the plot involving Karin, Mark, and Weber, the narrator muses on the sandhill cranes' annual staging along the North Platte River and their evolutionary history. In contrast to the certainty of the narrator that elucidates neurobiology, the voice in these sections is tentative and shows that knowledge of crane ethology is limited.[32] It's as if this narrator responds to Ressler's dread that, in the biotech labs of the future, "what can be filled in of the map will be filled in" by leaving room for the unfathomable and astounding.[33] Even as the lyrical narrator offers us information about migratory patterns, it reinforces limitations to understanding. Take the opening lines of part 2: "[The cranes] head for the tundra, peat bogs and muskegs, a remembered origin. . . . There must be symbols in the birds' heads, something that says *again*. They trace one single, continuous, repeating loop of plains, mountains, tundra, mountains, plains, desert, plains" (96, 98). The vague "something" and the searching but hopeful "there must be" intimate the cranes' mystique, and the lyrical narration compensates for its incomplete knowledge about crane migration by supplementing it with images "*to make the world strange again.*"

Alternating between the voice of information and the voice of verse, *The Echo Maker* also straddles the line between "accountability to matter and to discursive mentation," as Lawrence Buell puts it.[34] Like the nonfiction texts that Buell endorses in *The Environmental Imagination* (1995), Powers's novel answers to the "matter" that the researches of neuroscience and zoology help explain, but it also fashions images to help readers "see what without the aid of the imagination isn't likely to be seen at all."[35] Hence there is a place for both mimesis and invention in environmental writing, especially if we distinguish mimesis from bald imitation and favor Buell's thesis that, through stylization, mimesis can effect the "dislocation of ordinary perception."[36] By this account then, mimesis can promote Shklovskian defamiliarization, so long as prose experiment enlivens the presented matter. *The Echo Maker*'s opening performs "dislocation" in this way. "A neck stretches long; legs drape behind. Wings curl forward, the length of a man. Spread like fingers, primaries tip the bird into the wind's plane. The blood-red head bows and the wings sweep together, a cloaked priest giving benediction. Tail cups and belly buckles, surprised by the upsurge of ground. Legs kick out, their backward knees flapping like broken landing gear" (3). Depicting a crane's landing, these lines take the reader on a flight through metaphor as they render the mechanics of flight. The unfocalized narrator first anthropomorphizes the bird—feathers as "fingers," wings the height of a man—but concludes by mechanizing its anatomy.[37] Clipped sentences draw the reader's attention to the very act of attending: the narratorial eye takes in whole patterns in the preceding paragraph but settles here on the anatomy of the single creature. The impersonal narrator promotes wonder by crossing referential registers. "Primaries" is an ornithological term for the large feathers required for flight, but the phrase "blood-red" that appears in the subsequent sentence calls up the word's chromatic meanings. The text similarly activates the layered connotations, geometric and aeronautical, of "plane." The ambiguity of "primaries" and "plane" is not settled. That is because it and the linguistic acrobatics on display in this cropped paragraph are designed to go beyond information and inspire wonder at seemingly prosaic animal behavior and to kindle the revived vision that Karin later models.

The oscillation between familiarity and surprise that constitutes the affect of wonder is crucial to both the passage describing bird flight and Karin's experience at the Refuge. As Philip Fisher explains in his philosophical treatise on the emotion, wonder arises when we sense the "radical singularity" of an object that still delivers the "surprise of intelligibility."[38] The object cannot be so drastically singular that the mind fails to find a kernel of familiarity within it. For this reason, Fisher proposes that the wondrous occupies a middle place between the predictable and the irrational, between perception and intellection, and between aesthetics and science. It depends on a balanced ratio: the wondrous phenomenon takes one out of the ordinariness of everyday events— a swath of rainbow, bird migration—which at once makes one's surroundings surprising and, in a second move, reconfirms one's place in the world and ability to apprehend it. Succinctly put, "the relation of certainty to surprise . . . is the relation of wonder."[39] Certainty and surprise, familiarity and strangeness: this dyad is the motor of wonder, and it governs *The Echo Maker*'s dual narrative structure and style.

Powers constructs his text to prime readers for the affective relation of wonder. That is, the modernist passages that defamiliarize Kearney's riparian ecosystem train readers for the task of interpreting Karin's epiphany four hundred pages on. The lyrical narration instills the notion that the natural world is eternal yet incomprehensible, unpresentable save through metaphor, and autonomous from human dictates. A passage that renders the bioregion and the novelistic itself strange establishes the timelessness and opacity of the nonhuman, and it is particularly remarkable because it interrupts the main diegesis:

> A flock of birds, each one burning. Stars swoop down to bullets. Hot red specks take flesh, nest there, a body part, part body.
> Lasts forever: no change to measure.
> Flock of fiery cinder. When gray pain of them thins, then always water. Flattest width so slow it fails as liquid. Nothing in the end but flow. Nextless stream, lowest thing above knowing. A thing itself the cold and so can't feel it. . . .
> Not even river, not even *wet brown slow west*, no now or then except in now and then rising. (10)

Because these lines immediately follow Karin's first visit with Mark in his coma and don't preface a new part of the novel, we at first read them as the stream of her brother's damaged consciousness. Here are the marks of the modernist style: ellipsis, fragment, chiasmus ("a body part, part body"), and repetition ("no now and then except in now and then rising"). The narrator insists on the eternality and indifference of the river's flow, indifferent even to its human assigned properties of "*west*" and "*brown*." The North Platte is "not even river" because it defies its liquidity and incarnates cold. Such revitalized images of the more-than-human world largely constitute the novel's environmental imagination until questions about the ecosystem's viability preoccupy the story in part 4. Powers builds *The Echo Maker* thus strategically. When Karin expresses the need for "*something to wake sleepwalkers and make the world strange again*," readers call up the passages that depict a wondrous environment to which we do not have complete access but that is alluringly familiar. The text eschews both Daniel Riegel's environmentalism of "guilt and facts" and Robert Karsh's greenwashing through "unlimited budgets, sophistication, subliminal seduction" and entices the reader to environmental accountability through wonder (346). With this design, *The Echo Maker* makes the case that wonder, the affective companion to defamiliarization techniques, is a means of attaining environmental awareness, a case that depends on readers themselves alternating between lyric and exposition.

Wonder is not only the affect of defamiliarization; its workings are homologous to the model of cognition that *The Echo Maker* proposes. Pursuing this homology, I complicate my claim that the novel accords with *Gold Bug* and fits into the canon of works that mobilize wonder as a trusty conduit to environmental commitment. It is the affective hinge of ecosickness because it articulates environmental consciousness in terms of the human body in a state of disorder. Ultimately, however, the novel's reflections on those dysfunctions reveal the flip side of wonder and challenge this affect's role in paving the path to care. Wonder just might indicate a "readiness to find what surrounds us strange and odd; a certain determination to throw off familiar ways of thought," as Michel Foucault attests, but does it always provoke us to then don the mantle of involvement?[40]

"THE ORDINARY BY ANOTHER NAME"

In *The Echo Maker*'s neurological plot, wonder forges the link between ecological relation and brain function. Oscillation between the familiar and the strange, the mechanism that is so essential to wonder, also governs sight and perceptions of place and one's own body.[41] Taking apart this mechanism also uncovers the Janus face of wonder. In addition to "draw[ing] us into sustained rapport with—and action on behalf of—the wider environment," oscillations between strangeness and familiarity induce states that are inimical to care.[42]

Capgras syndrome brings to life a nightmare: everything you love becoming emotionally foreign. According to philosopher William Hirstein and neurologist V. S. Ramachandran's description of the disorder, the afflicted person no longer feels the emotion that he knows should attach to a loved one, and "the only way the patient can make sense of the absence of this emotional arousal is to form the belief that the person he is looking at is an imposter."[43] Despite the person's ability to recognize another's physical appearance, he or she cannot trust in that one's "psychological identity."[44] Capgras, therefore, exposes the instability of the "logic of consciousness" because it reprioritizes emotion and knowledge and substitutes false stories for "emotional arousal."[45] Through the delusional misidentifications that characterize Capgras, *The Echo Maker* advances the notion that unimpaired brains also function by smoothing over inconsistencies and maintaining a mental status quo. At an unnamed university in Stony Brook, New York, Weber teaches that the mind's main objective is precisely this, to ensure consistency at all costs: "'The job of consciousness is to make sure that all of the distributed modules of the brain seem integrated. That we always seem familiar to ourselves. . . . We think of ourselves as a unified, sovereign nation. Neurology suggests that we are a blind head of state, barricaded in the presidential suite, listening only to handpicked advisors as the country reels through ad hoc mobilizations'" (363). The brain is an isolated, imperious leader, likened here to George W. Bush planning his post-9/11 invasions of Afghanistan and Iraq. In Weber's lesson, the brains of individuals with Capgras-type neural damage continue their primary task of ironing out the wrinkles of thought, heeding only "handpicked advisors" despite contradictory input from the wider world. Capgras is an irresistible case for Weber because the brain insists on self-familiarity in the absence of a familiar attachment to beloved objects. In other words, the mind believes itself even though it no longer believes in those it would normally trust.

Weber demystifies the idea that only the neurologically ill must accommodate mental fragmentation when he recalls Sarah M., a patient with "a rare, near-complete motion blindness" called akinetopsia. "Sarah's world had fallen under a perpetual strobe light. She couldn't see things move. Life appeared to her as a series of still photographs, connected only by ghostly motion trails. . . . And yet, strangest of all: Sarah M. alone of all the world saw a kind of truth about sight, hidden from normal eyes. If vision depends upon the discrete flash of neurons, then there is no continuous motion, however fast the switches, except in some trick of mental smoothing" (107).[46] Her brain cannot carry out the neural process that welds the discrete units of vision into a seamless whole. A region-specific brain injury renders Sarah M.'s world radically strange, but this strangeness in fact accentuates the mechanism of "normal" sight. *The Echo Maker* thus interpellates its audience into a community of dysfunction.[47] Though Capgras is the clinically diagnosable disorder on which the novel centers, Sarah M.'s case shows that even baseline function exhibits the structure of so-called sickness. In other words, perception is impaired in that it depends on a "trick" of negotiating the familiar and the strange. In *The Echo Maker*, then, toggling between these poles is the form that ecosickness takes, but it is a sickness shared by all.[48] The neurological processes of perception and cognition and the affective structure of wonder are homologous, and they together govern environmental consciousness.

Wonder is not only central to environmentality because it enchants the subject's world. The novel proposes that the interplay between familiarity and strangeness that constitutes this emotion and visual processing also structures place sense.[49] This dyad thus governs how *The Echo Maker* figures the body and the individual's experience of his surroundings. The correspondence between these domains comes into focus as Weber travels from New York to Kearney and continues mulling Sarah M.'s case: "She was there, in Weber's strobing mind, when he stepped into the jetway at LaGuardia, and gone when he found himself, that same afternoon, dead center in the evacuated prairie, with no transition but a jump cut" (107). The narrative transposes the operations of sight onto those of thought through the image of the doctor's "strobing" mind and onto those of movement as Weber hops from city to country. The photographic and cinematic metaphors that figure akinetopsia—"still," "motion trails"—also characterize the disorienting experience of travel—"jump cut." The flight carries him from the stability of professorial life in New York to the turmoil of intertwined medical, familial, and sexual dramas. But it is not only the abrupt transition that disturbs the traveler. Movement itself is a haunting. Traces of the "ghostly" accompany us as we enter new surroundings, which we experience as "motion trails." Later in *The Echo Maker*, the postcard, an artifact of travel, reinforces this point:

> Somewhere . . . Weber described discovering a largely intact and responsive hand blossoming across the face of an amputee, Lionel D. Touched high up on the cheekbone, Lionel felt it in his missing thumb. Grazed on the chin, he felt it in his pinkie. Splashing his face with water, he felt liquid trickle down his vanished hand.
>
> Weber shut off the shower and closed his eyes. For a few more seconds, warm tributaries continued to stream down his back. Even the intact body was itself a phantom, rigged up by neurons as a ready scaffold. The body was the only home we had, and even it was more a postcard than a place. We did not live in muscles and joints and sinews; we lived in the thought and image and memory of them. No direct sensation, only rumors and unreliable reports. (260)

The body, which is itself an epiphenomenon of the true stuff of consciousness, is both as familiar as home and as unfamiliar as a far-flung destination. Remnants of the past help smooth over perceived changes in our surroundings to create a feeling that is as fluid as the tributaries of water that wet Weber's body. The same process, that is, the negotiation of familiarity and strangeness, governs both perception of the body (which the narrator figures *as* a place) and perception of the places one inhabits. *The Echo Maker*'s observations about travel externalize the mind's fluctuations between the unexpected and the readily recognizable as it perceives its surroundings.

Once Weber has landed in Nebraska, the built environment activates the play between familiarity and strangeness that inspires *The*

Echo Maker's accounts of wonder and cognition of body and place. Upon arriving in Kearney, Weber selects a hotel for its alluringly ambiguous sign, "welcome crane peepers" (107). He learns that it refers to Buffalo County's standout feature: the staging ground to which thousands of sandhill cranes migrate every February and March. The spectacular attraction that the sign announces belies the motel's more prosaic interior: "once inside the MotoRest lobby, he might have been anywhere. Pittsburgh, Santa Fe, Addis Ababa: the comforting neutral pastels of global commuting" (108). In other words, Weber finds himself in a "non-place" that, in anthropologist Marc Augé's account, is at once disturbingly recognizable and affectively neutral. (In this respect, the apperception of non-places is akin to Capgras.) Like the corporate hotels and transit hubs that Augé houses with his concept, the infrastructure that grows to support crane tourism has an eerie familiarity. Non-places "produc[e] effects of recognition. . . . A foreigner lost in a country he does not know (a 'passing stranger') can feel at home there only in the anonymity of motorways, service stations, big stores and hotel chains."[50] Powers's narrative, focalized through Weber, iterates Augé's observations: "the usual gamut of franchises—motel, gas, convenience store, and fast food—reassured the accidental pilgrim that he was somewhere just like anywhere. Progress would at last render every place terminally familiar" (166). Cookie cutter development along Interstate 80 threatens to efface the signature of this Midwestern place. The blandness of "a diffident commercial strip marked by a forest of metal sequoias bearing harsh, cheery signage" sharply contrasts Weber's initial perception of the plains as a land so foreign he needs a passport to gain entry (166). Similarly, the featureless amenities of crane tourism contrast the striking red crests of the birds that adorn tourist brochures in the MotoRest, and these stale amenities threaten to diminish the very water and land resources that support the cranes and maintain the strange singularity of the migration.

Weber's movement between Kearney's organic and built environments triggers cognitive transit between familiarity and strangeness. The cranes' behavior and history expand on this dynamic. In the oneiric depictions of the sandhill cranes that preface the narrative of Mark's recovery, *The Echo Maker* indicates a deep reason why these two sensations infuse place just as they guide cognition and constitute wonder. The persistence of bird migration gives Buffalo County an air of familiarity, while the magnificence of the birds' staging patterns astounds the viewer. The novel's opening captures the effects arising from the cranes' perennial choreography: "Cranes keep landing as night falls. Ribbons of them roll down, slack against the sky. They float in from all compass points, in kettles of a dozen, dropping with the dusk. Scores of *Grus canadensis* settle on the thawing river. They gather on the island flats, grazing, beating their wings, trumpeting: the advance wave of a mass evacuation. More birds land by the minute, the air red with calls" (3). The narrator breaks out of the temporality that dominates the rest of the text, switching from the past tense to present narration. The present progressive tense of the excerpt's first and last lines amplifies this temporality and "keep" reinforces it.[51] "This year's flight has always been," the introduction continues. "Something in the birds retraces a route laid down centuries before their parents showed it to them. And each crane recalls the route still to come" (4). A crane is a living, kinetic record of the instincts the species has accrued over millennia; therefore, its past and future suffuse its present. Fully encoded in the birds, this record is partially illegible to the voice that narrates these passages. The birds, centuries-old visitors to the North Platte, continue to strike spectators as alien. The inevitability of their migration's "continuous, repeating loop" yields surprise, marked by the synesthetic phrase "air red with calls," and, as I noted earlier, the narrator's uncertainty (98).

The patterns of human cognition harmonize with the patterns of crane movement, yet *The Echo Maker* refrains from idealizing human attachment to place. Rather, the text contrasts humans and cranes based on whether their environments are salutary, a contrast that pivots on geographical determinism.[52] The narrative counterposes the birds' irrepressible, awe-inspiring fit with the bioregion to Karin and Mark's unwilled submission to a home that damages them. Soon after the text relates the cranes' embeddedness in Nebraska's evolutionary fabric, Karin defines home as "the place you never escape, even in nightmare" (8). She focalizes a description of Kearney's downtown core that recasts the idea of a "remembered origin" as an unshakable yoke: "She cut through downtown Kearney, a business district hosed for as far into the future as anyone could see. Falling commodities prices, rising unemployment, aging population, youth flight, family farms selling out to agribusiness for dirt and change: geography had decided Mark's fate long before his birth. Only the doomed stayed on to collect" (28). While the town is an agent insofar as residents stagnate under its influence, it is powerless to resist the economic changes that are leaving it behind. Midwesterners can no longer collect on the promise that the land will nourish their future. Karin's view of geographical influence under centralized industrial agriculture inverts the idealism of early American agrarians such as Thomas Jefferson and writer-farmer Hector St. John de Crèvecoeur. Leo Marx expounds on the latter's assertions that "men everywhere are like plants, deriving their 'flavor' from the soil in which they grow. In America, with its paucity of established institutions, however, the relation between mankind and the physical environment is more than usually decisive. . . . At bottom it determines everything about the new kind of man being formed in the New World."[53]

Over two hundred years after the publication of Crèvecoeur's *Letters from an American Farmer* (1782), *The Echo Maker* still holds that the environment sculpts the character of its inhabitants. However, the earth no longer has a "salubrious effluvia" that produces a "new kind" of cultivating and cultivated man.[54] Instead, Karin fingers the place itself as a blight on the new crop of settlers. Like other Generation X Nebraskans, Mark is a farmer manqué without a land inheritance. His mental disintegration expresses the decay of the earth signaled by the growth of agribusiness (the Iowa Beef Processors plant that employs Mark and his friends)[55] and tourism (the Great Platte River Road Archway Monument, "'the only monument in the whole world that straddles an interstate'" [38–39]).[56] The novel fictionalizes the real erosion of the family farm due to the concentration of corporate agricultural practices that decimate profits for small-scale farming and lead younger generations to abandon farm life in bankruptcy or fear thereof.[57] Depicting the decline of independent agriculture in the U.S., the novel does not thereby idealize ties to the land. Rather, environmental determinism deromanticizes the notion that the land as currently sculpted by agribusiness and tourism can any longer improve somatic and socioeconomic health.

Across its disparate narrative lines—from a man's misidentification disorder to Midwesterners' alienation from a salutary home to cyclical bird migrations—*The Echo Maker* envisions a disturbing form of estrangement: estrangement from the most familiar. Taking apart the novel's affective and perceptual motors—that is, the toggling between familiarity and strangeness that comprises wonder and mental processes—shows how wonder affiliates the novel's environmental and medical plots around their affective continuity rather than causality. This correlation does not signal the harmonious synthesis of human and environment, however; it recasts environmentalist accounts of ecological awareness. The rest of this chapter establishes that, while seeing the ordinary as strange can, as Thomashow writes, sink you into "the place from which you originate" and inspire reverence,[58] the everyday defamiliarizing processes of consciousness threaten the conversion of that disposition into an ethics of care. Environmental thinkers might promise that, through "a place-based perceptual ecology . . . you learn how to pay closer attention to the full splendor of the biosphere as it is revealed to you in the local ecosystem," but the lessons of neuroscience teach that nothing is "revealed to you" in its pure state.[59] As *The Echo Maker* expounds two other tropes common to environmental thought, connection and care, it posits that there may be "no direct sensation, only rumors and unreliable reports" (260).

"STRUGGLING WITH COMPLEX INTERACTIONS"

As a new addition to the family of nature writers descending from Thoreau, *The Echo Maker* picks up on the tension between connection and disconnection that animates this literature. The novel validates the claim that thinking nature often entails thinking the mind. That is, as Sharon Cameron observes, "to write about nature is to write about how the mind sees nature, and sometimes about how the mind sees itself."[60] In Powers's novel, these two inquiries interdigitate as the text makes wonder the emotional fulcrum of environmental engagement. *The Echo Maker* diverges from the nature writing tradition, however, by suggesting that this engagement depends on the picture of the mind that one draws. It pins the potential upset of environmental involvement on the very toggling between familiarity and strangeness out of which wonder and connection making arise. In the final analysis, a sense of connectedness fails to solve the difficulty of getting outside of the self to care for others. This is because systemic and affective complexity proliferate as the vectors of connection multiply and produce negative emotional states.

In a 1999 interview, Powers defines the novel as "a supreme connection machine—the most complex artifact of networking that we've ever developed."[61] Four years later, he reasserts, "Ultimately my books are about connection. They turn upon the truth that there is no independent mode of existence."[62] Powers's fascination with systems that exhibit dependent complexity is apparent in the foci of *Galatea 2.2* (neural nets), *Gain* (the tentacular reach of a multinational corporation), *Plowing the Dark* (virtual reality programming), and *Generosity* (the neuroscience of altruism). *The Echo Maker* extends the connection trope into neurological and ecological domains and envisions interdependence occurring on three levels: within the neural system, between humans, and between species in an ecosystem. As it theorizes the relays between brain, self, and human and nonhuman other, the novel argues that the flicker of familiarity and strangeness models the ebb and flow of connectedness and that this movement generates the paranoia and projection that can jam one's ability to connect with others.

Weber disseminates *The Echo Maker*'s view of neural interconnectedness. As we have already seen, his holistic account of brain physiology emphasizes that the brain cobbles together wholeness out of fragmentation. "We were not one, continuous, indivisible whole," Weber insists in his book, *The Country of Surprise*, "but instead, hundreds of separate subsystems, with changes in any one sufficient to disperse the provisional confederation into unrecognizable new countries" (171). Take vision as an example. It "requires careful coordination between thirty-two or more separate brain modules. . . . Only: the many, delicate hardwires between modules can break at several different spots" (149). According to this modular view, dependent connectivity is integral to the system's functioning but makes it especially vulnerable to collapse.[63] Weber goes against the functionalist tide of neuroscientific research that, in the early 2000s, is locating the seat of personality traits, abilities, and beliefs in an isolated brain area. In his first conversation with Chris Hayes, Mark's neurologist in Kearney, Weber challenges the novice doctor's conviction that Capgras is purely neuronal:

> "There's a higher-order component to all this, too. Whatever lesions [Mark] has suffered, he's also producing psychodynamic responses to trauma. Capgras may not be caused so much by the lesion per se as by large-scale psychological reactions to the disorientation.... Clearly Mark Schluter's Capgras isn't primarily psychiatric. But his brain is struggling with complex interactions. We owe him more than a simple, one-way, functionalist, causal model."...
>
> The neurologist [Hayes] tapped the film on his light box. "All I know is what happened to his brain early on the morning of February 20." (132–33)

Hayes's gesture reveals his myopia: he limits the area of concern to the scans bounded by the light box, ignoring the interactions between different components of the neural system. Despite the technical sophistication of neurological research and practice, Hayes's view is ultimately regressive. He rides what Ian Hacking has dubbed the "neo-Cartesian" wave in biomedicine "whereby bodies are just machines in space, composed of machine parts, while the mind, the soul, is something else."[64] We have to adjust this model for brain science. In neo-Cartesian neuroscience, the *mind* is a brain composed of manipulable components, and function and dysfunction depend on the workings of these components or brain "regions." They turn on or off, dim or glare, to the benefit or detriment of vital functions like memory, fear, and recognition. As Lisa Zunshine notes, functionalism is ultimately part of a quest for essences that, she argues, in a seemingly essentializing move, results from "the quirks of our cognitive architecture."[65] Weber adheres to a different model: he is a holist.[66] Holists look behind the seemingly direct relationship between a perturbation to a system and its response at the intermediate, stochastic events between cause and effect. While a specific region of the brain certainly endured material damage, the system's response to the injury is likely a more proximal cause to Capgras than the injury itself. Narrating the homologies between mental processes and relations to human and more-than-human others, *The Echo Maker* shares Weber's holism. Injury and response must be approached in their relation to each other, to their local and wider environments, and to the people that become enwrapped in them. Stories are vectors for this understanding because they accommodate multidimensionality. They "narrate disaster back into livable sense" as they help us tell "ourselves backward into diagnosis and forward into treatment" (414).[67] As will be clear in my later reading of *The Echo Maker*'s ending, Weber's faith in storytelling reaches its limits when stories confront the "disaster" of environmental threats and the patterns of human behavior that lead to them.

Weber's holistic stance is of a piece with the work of neurophysiologist Giacomo Rizzolatti and philosopher Corrado Sinigaglia on mirror neurons. Their research postulates that mirror neurons "allow our brain to match the movements we observe [others perform] to the movements we ourselves can perform, and so to appreciate their meaning. Without a mirror mechanism we would still have our sensory representation of the behavior of others, but we would not know what they were really doing."[68] Humans and other species of animal are able to move and speak because they have encoded another's prior performance of these behaviors. Rizzolatti and Sinigaglia conclude from this that mirror neurons are "at the root of our capacity to act as individuals but also as members of a society" and that connection is indispensable for awareness of our social being.[69] They conclude the preface to their study with the fundamental claim that mirror cells "show how strong and deeply rooted is the bond that ties us to others, or in other words, how bizarre it would be to conceive of an *I* without an *us*."[70] They are optimistic that understanding mirror neurons can foster empathy within human communities, an expectation that Weber momentarily shares when he muses that greater awareness of our shared mental deficits might enhance identification with others (234). Yet, when *The Echo Maker* portrays the ramifications of Capgras and interdependence across systems, it exhibits less confidence that connection is always so ethically and socially salutary.

In *The Echo Maker*'s calculus of connection and disconnection, contagion is the negative of the mirror neuron concept and invites skepticism about the upshot of interdependence. Several months after Mark's accident, his rage at the "fake" Karin intensifies, and Karin acknowledges the toll of his sustained doubt: "The Capgras was changing her, too. She fought against her habituation. . . . For a little while longer, she knew the accident had blown them both away, and all the selfless attention from her in the world would never get them back. There was no *back* to get them to" (236–37). Proximity to her brother's disease recalibrates Karin's brain. Due to her resolute attention to Mark, she "catches" an inverted version of Capgras through which she sees her post-injury brother as more truly *Mark* than the insistent memory of the original. Weber also integrates himself into Mark's disorder, not only as healer but also as likely cause for its continuation. The longer that he entertains Mark's explanations for his disorientation—that he's been "programmed in a government machine" or has "been living in a video game," for example—the more seriously Weber weighs whether "maybe he'd been helping the man create this illness. Iatrogenic. Collaboration between doctor and patient" (303).[71] Of course Weber exaggerates his influence on a condition that antedates his arrival, but his egoistic reflections highlight the novel's argument that the neural system generates feedback, which then encompasses entities that are ostensibly external to it. Not only is Mark's brain "producing psychodynamic responses to trauma" and "large-scale psychological reactions to the disorientation," his caretakers' "infected" minds are as well (132). Connection is certainly "at the root of our capacity to act as individuals [and] also as members of a society," as Rizzolatti and Sinigaglia hold.[72] However, *The Echo Maker* pushes against the idea that connection is mutually beneficial and instead reports the potential harms of connection that intrude on the practice of care.

The Echo Maker therefore puts to the test the central tenet of the ecological perspective on connection that the novel advances through Daniel Riegel. In part 1 the narrator introduces Daniel as a magnanimous ascetic who lives in a "dark monk's cell" and supports Karin even though she had cheated on him six years ago (70). Karin focalizes the first, ambivalent descriptions of Daniel when she shares with him her admiration for Barbara Gillespie, Mark's overqualified rehabilitation nurse.[73] Barbara nurtures her patient with total ease, and Karin considers whether fate has brought the nurse to the Schluters.

As she relates this idea, "Daniel stood and walked to the window, stark naked, oblivious. Like a wild child. The chill of his apartment didn't touch him. He tried on the idea. She loved that in him, his eternal willingness to try her on. 'No one is on a separate path. Everything connects. His life, ours, his friends'. . . . mine" (72, ellipsis in original). Daniel assures Karin that Barbara and Mark are indeed entangled in a single web of relations when he delivers lines that chime with the first of Barry Commoner's "laws of ecology": "Everything is Connected to Everything Else." This law proclaims "the existence of the elaborate network of interconnections in the ecosphere: among different living organisms, and between populations, species, and individual organisms and their physico-chemical surroundings."[74] Evoking Commoner, Daniel recites an ecological perspective that, as we saw in chapter 2, resounds in U.S. environmentalist discourse.

Karin may believe this theory of interconnection, but the habit of *living* relatedness exceeds her. During a rare argument with Daniel, "pointlessness flooded her, the futility of all exchange. . . . She felt a deep need to break everything that pretended to connection. . . . Love was not the antidote to Capgras. Love was a form of it, making and denying others, at random" (268). Karin troubles Commoner's law through the language of fakery: if humans "pretend to connection" and "mak[e]" others, connection is in fact egoistic. These phrases intimate that what Daniel perceives as the bond between humans and nonhumans is merely projected longing. Through Karin's eyes, "everything is connected to everything else" to the extent that thinking and desire make it so.

Such give-and-take on connection also occurs at the level of metaphor. Powers tempts readers to idealize connection by weaving the ecological perspective into *The Echo Maker*'s tapestry of images, but the color of those images hints at the violence of connection and the vulnerabilities that it engenders. The first thread of this fabric appears early in the book when the narrator describes Mark after his operation: "His head was shaved, with two great riverbeds scarring the patchy watershed. His face, still scabbed over, looked like a ten-inch peach pit" (38). Mark becomes Kearney's riverine environment via description, but this "scarring" river is grotesque and divisive rather than serene and harmonious. In a TV interview, Weber reads a passage from *The Country of Surprise* in which a cognate environmental simile explains the operations taking place inside the head. "We're more like coral reefs. Complex but fragile ecosystems . . ." (186). An injurious topography (scarring rivers) figures Mark's surface wounds while internal neurological injuries prompt an image of vulnerability (coral reefs) to explain consciousness. Ultimately, *The Echo Maker* exhibits a feature common to the ecosickness narratives of Jan Zita Grover, David Foster Wallace, and Leslie Marmon Silko: damage or susceptibility to damage taints the metaphors that set up the kinship between the human and the more-than-human. Like the familiarity and strangeness dyad, these topographical metaphors establish the pervasiveness of ecosickness, homologize somatic and environmental damage, and recast connection as potentially harmful.

Through its dueling positions on connection—on the one hand, Weber's holistic account of the brain, Daniel's ecological perspective, and the metaphorization of the human through natural topography; on the other hand, Karin's skepticism and the negative tone of those metaphors—*The Echo Maker* intervenes in humanist observations about ecological science. By recoding ecological principles such as connection, the text follows a practice proposed by Dana Phillips, who admonishes cultural critics "to realize that ecology is not a slush fund of fact, value, and metaphor, but a less than fully coherent field with a very checkered past and uncertain future."[75] Environmentally oriented authors and critics have "seized upon ecology as an accessory and complement to their own brand of professional discourse because of their commitment to environmentalism, and because they have thought that ecology offers scope for the vibrant depiction of a natural world conceived of organically."[76] Yet they have ignored debates within and outside of ecological study over the veracity of contested concepts such as harmony and homeostasis.[77] As chapter 2 detailed, when we challenge these concepts we also must rethink how we represent environmental relation and promote environmental care. Ecology does not provide a stable vehicle for equating textual operations to environmental ones. Portraying an environment and a mind out of balance, *The Echo Maker* admits the appeal of the sentimental view of ecological harmony and invokes connection as a motif for expressing environmental investment that has been expedient for at least the last two centuries. However, in detailing cognitive processes the narrative uncovers the limitations to connectedness and bolsters Phillips's case against unreflective use of ecological metaphors. The text acknowledges that interconnection is required for different types of systems to function, but, as it depicts the emotional entailments of oscillating between familiarity and strangeness within the brain and across the self-other boundary, the novel shows those connections to be friable and dangerous.

Karin indicates one reason why connection misfires to obstruct true care. Musing on developer Robert Karsh's influence over the public, she concedes that "we love only what we can see ourselves in" and invent what we see based on our own interests (343). Ultimately, associating connection with projection, *The Echo Maker* ponders whether seeing connections is another way of thinking that everything is about you. The condition of paranoia, which is a symptom of Mark's Capgras, and the affect of paranoia, which spreads to those proximate to him, also point to the ugly obverse of the connection making that had seemed so generative in wonder. Eve Sedgwick elaborates the structure of paranoia in terms instructive for understanding this disposition in Powers's text. Paranoia is relational. It "tends to be contagious. More specifically, paranoia is drawn toward and tends to construct symmetrical relations."[78] In this respect, it attests to the peculiar transmissibility of affect, its capacity "to enter into another."[79] Paranoia is also mimetic. As I observe the paranoiac's absolute doubt, I absorb and imitate his stance toward the world.[80] In these ways, the attributes of paranoia repeat the attributes of connectedness that *The Echo Maker* expounds. In addition to being mimetic, the paranoia of Capgras produces fear of imitation as one of its symptoms. Mark imagines that Karin and his dog are copies, and, as his disease mutates, he becomes convinced that Daniel is reproducing himself and stalking Mark. Generating and projecting (false) familiarities, Mark's brain keeps experiences of the new and strange at bay. That is, it feeds any information Karin advances to rebut his accusations back into the story that his brain has constructed since the injury. Unlike Grover and Wojnarowicz, whose suspicion produces the new in their AIDS memoirs, Mark is taken over by an indomitable paranoia; his brain scrambles the input that would reconfigure his convictions and introduce surprise.

Paranoia, therefore, makes everything excessively familiar, but to the point of alienating the individual from all others. As Mark's condition evolves to include paranoia along with Capgras, access to productive wonder is blocked because his mind enfolds the potentially strange into his habituated patterns of paranoid thought.

While explicating complexity, a task that has alternately attracted and irritated critics of Powers's work, *The Echo Maker* generates friction between ecological interconnectedness and egoistic or paranoid forms of connection making.[81] The novel adopts the ecological perspective that "we're all connected" as the ready explanation for neurological processes and the interactions between humans and the environment, but it hits on the snags of connection when it elaborates projection and paranoia. Advancing a tradition of environmental thought, *The Echo Maker* imagines that the wondrous oscillation between the familiar and the strange primes us to engage with and care for the nonself, but this mechanism can tip us into affective positions that cut us off from others. Unlike affects such as disgust, which in Wallace's *Infinite Jest* balances closeness to and detachment from the outside world, wonder, gone wrong, blocks attachment. It shades into projection as we make the other conform to our own desires and into paranoia as the strange becomes excessively familiar. These states then seal off openness to the outside world. Powers's novel introduces the possibility that connection can generate two opposing effects on self-other dynamics: connectedness brings us into relation with our fellow humans and the more-than-human realm and thereby develops our minds, and yet connection takes us further into ourselves. *Gold Bug* concludes with Jan's revelation, "nothing deserved wonder so much as our capacity to feel it," that wondering at our own wonder might just put care before cold information.[82] *The Echo Maker* takes up the task of palpating human wonder, but the connections on which wonder depends become a problem for care rather than its helpmate. By cautioning readers about the insulating aspects of connection making, *The Echo Maker* contributes to debates on whether environmental discourse too readily trusts that a sense of interdependence cures apathy and detachment and posits that connection can divert rather than direct care.

"THE ETHIC OF TENDING"

Can an ethic of care gain traction in the unstable movement between familiarity and strangeness, between wonder and paranoia, between connection and disconnection? The call to care motivates action in *The Echo Maker*'s medical and environmental narratives, and, in both plots, the tensions of these dialectics govern a person's ability to tend to the external world. Even a novice reader will note that Powers's novels detail the promises and pitfalls of rampant technoscientific development. Less obvious from a cursory read is how often such patently cerebral texts reflect on care for the other. His works mull over this knotty question: Does care threaten personhood or constitute it? Across his oeuvre, Powers places this question at the center of narratives on "the ethic of tending."[83]

It is this ethic that concerns Jan O'Deigh as she immerses herself in the history of genetics research and "learn[s] that I live in an evening when all ethics has been shocked by the sudden realization of accident."[84] The erstwhile librarian "feel[s] sick beyond debilitation to think what will come, how much more desperate the ethic of tending is, now that we know that the whole exploding catalog rests on inanimate, chance self-ignition.... We've all but destroyed what once seemed carefully designed for our dominion. Left with a diminished, far more miraculous place—banyans, bivalves, blue whales, all from base pairs—what hope is there that heart can evolve, beat to it, keep it beating?"[85] The destruction Jan mentions is twofold: genetic research has eliminated the idea that the human species is an inevitable outcome of an intentionalist plan and humans have eliminated their companions on the planet. Jan wonders how the discovery that all of life reduces to the base pairs of DNA will translate into a care ethic rather than solipsism and the destruction of the more-than-human. *Gold Bug* presents a clear trade-off, but one without an easy resolution: care takes us out of the solipsism that genetic discoveries seem to induce, but giving the self over in care threatens individual autonomy. One of the novel's minor characters, Annie Martens, exhibits a "great capacity for care," but "the need to distribute surplus care led her to sacrifice personal preference to subscribed taste.... Nothing mattered except giving compassion in the available dialect."[86] The challenges of care, then, add another dyad to those that this chapter has explored thus far, and one that will interest Wallace in *Infinite Jest*: the pull between autonomy and the relinquishment of selfhood.

Karin's vexed relationship to Daniel stages this problematic in *The Echo Maker*. When Daniel introduces her to Weber's work, she feels both grateful for and shackled by his munificence. "She was in Daniel's debt again," she regrets. "On top of everything else, he had given her this thread of possibility. . . . Of all the alien, damaged brain states this writing doctor described, none was as strange as care" (94). The focalization of this last sentence is noteworthy. We initially assume that Karin, whose point of view we move in and out of, is the source of this thought along with the others cited earlier. But because Karin has not yet read Weber's books at this point in the story, she does not yet have the neurological knowledge necessary to observe that no brain state is "as strange as care." A heterodiegetic narrator speaks this dictum to present the idea that care is an odd state because it requires at least a minimum of self-renunciation. Daniel's magnanimity, like Capgras and akinetopsia, is an improbable disorder that compromises autonomous selfhood. Even Daniel admits as much when describing his meditation practice: "'It makes me more . . . of an object to myself. Disidentified'" (43, ellipsis in original). Detached from himself *qua* self, Daniel relinquishes habits of mind. It "'makes my insides more transparent,'" he haltingly explains. "'Reduces resistance. Frees up my beliefs, so that every new idea, every new change isn't so much . . . like the death of me'" (43, ellipsis in original). Daniel here admits that connection is threatening if enlightening: when we confront the other and the "new ideas" that she puts before us, we risk self-annihilation. In light of the strangeness of care, Daniel's reparative selflessness appears even more aberrant.

If Daniel is a paragon of willed self-erasure, Mark exemplifies the breakdown of the will and figures the violability of the person and loss of control. He cannot orient himself in his now unfamiliar surroundings; moreover, he cannot control treatment of the primary injury and has no say in his therapy. Having lost his medical self-determination, Mark is no longer what Nikolas Rose terms a "biological citizen," someone who can "elaborate a set of techniques for managing" to optimize his somatic state.[87] In Mark's case, both health and personhood are under strain. For the Capgras patient, the state of being "disidentified" to which Daniel aspires amounts to near disintegration even as it forces him to rely fully on his own perceptions and to exclude others' intelligence. In a moment of desperation, Mark expresses this disintegration to Weber in the plea, "'Where's *me*?'" (415). Through Mark and the other cases that Weber recounts, *The Echo Maker* corroborates Susan Sontag's observation that "illness is the night-side of life, a more onerous citizenship. Everyone who is born holds dual citizenship in the kingdom of the well and in the kingdom of the sick. . . . Sooner or later each of us is obliged, at least for a spell, to identify ourselves as citizens of that other place."[88]

Through Capgras, Mark journeys across this border, from a "unified, sovereign nation" into territory trampled by "ad hoc mobilizations" (363). Along the way, he loses membership in the "kingdom of the well" and in the human kingdom, a transformation that takes place through description. Stepping up to Mark's hospital bed, Karin greets "a face cradled inside the tangle of tubes, swollen and rainbowed, coated in abrasions. His bloody lips and cheeks were flecked with embedded gravel. The matted hair gave way to a patch of bare skull sprouting wires. The forehead had been pressed to a hot grill. In a flimsy robin's-egg gown, her brother struggled to inhale" (7). Technomorphism alternates with zoomorphism in these early descriptions. Initially a deconstructed machine, Mark mutates into the cinematic version of Frankenstein's monster after "they slit his throat and put a bolt in his skull" (9). Karin catches a glimpse of him, but then "look[s] away, anywhere but at his animal eyes" (7). As Mark struggles to regain speech, the animal metaphors intensify: "The days of rehab drill numbed with crushing repetition. An orangutan would have started walking and talking, just to escape the torture. . . . Mark stared at her moving mouth, but wouldn't imitate. He just lay in bed murmuring, an animal trapped under a bushel, afraid that the speaking creatures might silence it for good" (34). Mark's cyborgism and animality stand in for his devastation and helplessness. Rather than substantiating Donna Haraway's influential thesis in "A Cyborg Manifesto" that the cyborg opens up a new (partial) subjectivity defined by the possibility of "*pleasure* in the confusion of boundaries,"[89] *The Echo Maker* here suggests that transgressing boundaries that delimit humanness can be severely disabling rather than enabling and generative.

Despite this disagreement, both Powers's novel and Haraway's manifesto contest the liberal ideology of self-possession. This ideology fractures in the novel, as it spins scenarios of ecosickness in which injury, on the one hand, and the desire to care, on the other, impinge on autonomy. The text demarcates the line between self-possession and its loss or sacrifice by examining freedom in light of the exigency of acting. Powers's books frequently puzzle over the urgency of action in the context of rampant technoscientific development and intervention. In *Gold Bug*, Stuart Ressler watches debates on the need for a nuclear test ban treaty and is torn between scientists' resolute search for information at all costs and a more humanitarian call for restraint. He eventually adopts the latter perspective, asking, "Aren't we graced with some degree of foresight? Is what we can do always what we must? . . . Until this moment, he was certain that the highest obligation of science was to describe objectively, to reveal the purpose-free domain. But here are [Edward] Teller and [Linus] Pauling carrying on on national TV as if some things were more urgent than truth, as if we're condemned always to fall back on the blind viewpoint of need."[90] Ressler's provocative questions raise a series of paradoxes. Experimental science, for some the apotheosis of human agency and self-determination, is "condemned" to continue on its unrelenting path to uncover truths. Willed action is ultimately compulsive. *The Echo Maker* puts these quandaries in the hands of the two doctors who spar over Mark's treatment plan. When Dr. Hayes rejects cognitive behavioral therapy, Weber reflects that, while an ideal treatment may not exist, "the first rule of medicine was to do *something*. Useful or worthless, however irrelevant or unlikely—act" (173).[91] This dictate is worrisome because it rhymes with the Hippocratic principle "to do good or to do no harm" even as it jettisons it.[92] In pursuit of care, the doctor's will supersedes the patient's best interests. Posing such questions over biomedical intervention, *The Echo Maker* reproduces the tangle of ethical concerns that grew up in *Gold Bug*: the pull between agency and self-restriction under the pressure to act. The later novel's dual narratives of sickness play out the vexing conundrum of whether humans can reasonably shackle the self in their efforts to care best for the other.

Karin's schooling in habitat decimation and resulting environmental awakening introduce this question into the text's environmental plot. *The Echo Maker* opens with an ecological reverie in a modernist, lyrical style that sparks environmental attention, but the urgency of environmental damage surfaces in didactic moments that elucidate Kearney's resource and land use disputes. Daniel reports, "'[Developers] have their eye on some parcels of land for a new project. Some open tract, near the river. We blocked them two years ago. Snatched four dozen acres out from underneath them. They're gearing up for war again . . . They're looking at riverfront. And whatever they build will increase usage. Every cup that comes out of that river reduces flow and encourages vegetative encroachment'" (265–66). Paradoxically, the erosion of the crane's habitat harms the birds even as it makes their perennial staging more spectacular and wondrous. "'The river slows; the trees and vegetation fill in,'" Daniel continues. "'The trees spook the cranes. They need the flats. . . . They used to roost along the whole Big Bend: a hundred and twenty miles or more. They're down to sixty, and shrinking. The same number of birds crammed into half the space. Disease, stress, anxiety. It's worse than Manhattan'" (57). Daniel anthropomorphizes the birds' reactions to overcrowding in order to heighten the need to protect their habitat. The patterns of water and land consumption indicate that the cranes cannot endure indefinitely, despite the lyrical narrator's amazement at their evolutionary persistence.

Karin stifles a laugh at Daniel's metaphor of cranes becoming neurotic city dwellers, but several months later she feels the seriousness of the matter. Stemming the river's overuse is now unignorable, and her job has become a vocation. For the first time since leaving social work to become a customer service representative at a computer company, Karin feels pulled to "larger service" and advocates for something beside herself and her kin (15). "What had begun as an invented job, the charity of a man who wanted to keep her nearby, had turned real. It was no longer even a question of meaningful work, of self-fulfillment. As absolutely delusional as it would have sounded to say aloud to anyone, she now knew: water wanted something from her" (398). And what water wants is for her to remain on her natal grounds and relinquish the idea of self that she had pursued in abandoning

Kearney. Spreading the Refuge's message is not egoistic or a step in an individualist project of upward mobility. Rather, her occupation has acquired the force of need at the expense of any plans for the self.[93] To grasp her newfound environmental engagement, Karin must anthropomorphize the environment—give it a voice that mutes her own—and hear it summon her to its aid. Karin correctly observes that hearing a call from water is fantastical; after all, what water really needs from her and from all humans is not to be needed. In fact, people have put the cranes and their habitat in the perverse position of relying on the benevolence of their own abusers. This irony strikes readers when the narrator sketches Daniel's hope for humanity: "Daniel needed humans to rise to their station: conscious and godlike, nature's one shot at knowing and preserving itself. Instead, the one aware animal in creation had torched the place" (57). Weber's and Karin's responses to somatic and ecological injury announce the necessity of taking action, but also draw out the positions of domination and practices of projection that come with connectedness.

Like sickness, care undermines the autonomy of the person, an autonomy that can also cut one off from the outside world. Yet, by figuring the debilitating effects of neurological disorder, *The Echo Maker* advises that we should not idealize the total relinquishment of autonomy. Neither a critique of liberal individualism, then, nor a call to self-renunciation, the novel engages in the same oscillatory thinking about care that it engages in when working through the familiarity-strangeness and connection-disconnection binaries. As characters follow an itinerary from perception of environmental and medical dilemmas to involvement in them, the narrative opens up the complexities that destabilize these positions. It testifies to what Robert Pogue Harrison sees as "the care-dominated nature of human beings,"[94] but shows environmental thinkers that care is not a "natural" outcome of wonder-filled attunement to complex connectivity.

The Echo Maker makes a strong case for wonder as an affect that sweeps the subject up in environmental attunement, revives awareness of somatic and ecological processes, and goads investment in problems of environmental degradation. On closer examination, the series of unresolved, interrelated tensions that structure perception and cognition, environmental relation, and the narrative itself muddy the transparency of awareness. While these defamiliarizing vacillations may produce wonder, they also introduce paranoia and projection and obfuscate the pathways between perception and care. To gather up this chapter's threads, I conclude on the question of "what next?" The affective mechanisms of *The Echo Maker*'s story of ecosickness raise questions not only of *what* we should do to care for human and nonhuman others (wanting to cure Mark is a given, just as protecting bird habitat is) but also of *how, to what ends*, and *at what cost*.

Undoubtedly, Daniel, Karin, and Barbara are all committed to preserving the crane's staging grounds, and yet all three pronounce the inevitable futility of their mission. With the novel's conclusion, readers discover that Barbara, Mark's nurse, came to Kearney as a journalist. Tasked with writing a fluff piece on crane tourism, she is drawn instead to the broiling water preservation and development issues. "'I dug a little,'" she tells. "'It didn't take much to find the water. I dug a little more. I learned that we were going to waste that river, no matter what I wrote. I could tell a story that broke people down and made them ache to change their lives, and it would make no difference. That water is already gone'" (435). In a minor metafictional moment, the text announces the foregone conclusion that growth and consumption will triumph over restraint and preservation. Barbara echoes Karin who, in the moment of her epiphany, acknowledges the "futility of her work" (407), given that the river is "down to gallons here. Hours and ounces" (446).[95] With this observation, the book closes with both Mark and the North Platte bioregion living on borrowed time. Mark's treatment with the antipsychotic drug olanzapine and mild electroshock therapy is working, and he now recognizes Karin. However, through prolepsis, the narrator warns that "in three months, her brother will be gone again, or his sister will, someplace the other won't be able to follow. But for a little while, now, they know each other" (447). *The Echo Maker* ends on a note of ephemerality that pointedly contrasts the eternality of its lyrical opening. Read side by side, the novel's introductory and concluding pages produce another contest: wonder competes with pessimism as its extreme form risks paranoia.[96]

The Echo Maker does not so much resolve the affective and ethical quandaries that ecosickness introduces as it plays them out to expose their facets. Rather than provide readers with emotional templates for fulfilling ethical imperatives, it probes the perceptual and affective adjustments that occur as individuals manage response to injury and threat.[97] What does it mean to awaken to the connections between self and human and more-than-human other at the same time as one awakens to the vulnerabilities and paranoia that connectedness introduces? How can a model of consciousness and an affect built around vacillation yield the minimum of stability required for action? In the final analysis, the motivating affect of Powers's "vision of wonder on a bleak Nebraska prairie" disrupts a tradition of quickening environmental consciousness through this inspiring emotion.[98] U.S. nature writers have trusted that, in Scott Slovic's words, "the kindling of consciousness—one's own and one's reader's—is a first step" to curing environmental ills.[99] Powers's contemporary environmental fiction confounds the next steps as it theorizes the constituent unreliability of consciousness and the workings of connection that upset care. Anchoring environmentality to cognitive processes, the novel performs the misfire of affect that it narrates through Mark's Capgras syndrome. *The Echo Maker* suggests that, when the writer "kindles consciousness," he cannot be certain what will end up aflame. These unexpected outcomes of narrative affect lead to the next chapters on Wallace and Silko. In *Infinite Jest* the aesthetic of disgust promotes attachment against endemic detachment, while the generalized anxiety of *Almanac of the Dead* stymies the revolutionary action that the novel chronicles and anticipates.

4

Infinite Jest's Environmental Case for Disgust

In 2008 WWF (formerly World Wildlife Fund for Nature) developed a campaign around two horrors of climate change: devolution and mutation. In the year when "change" became an unavoidable buzzword as part of Barack Obama's first presidential campaign, the environmental group revised the word's connotations to include devastation and unsettling alterations to the essence of life. Commissioned by WWF, Belgian advertising firm Germaine offered a commentary on the fate of humanness in the face of unabated climate change (figure 4.1). Adapting to rising sea levels and a new aquatic environment, humans have morphed into a chimerical life-form: a fishman. The bubbly pattern of the creature's shirt evokes the novel species' watery milieu. His mundane attire renders this radical change prosaic, but the ad's imperative copy, "Stop climate change before it changes you," shouts out how astonishing the figure is. The hum-animal's visage triggers unease. Green hues highlight its putrid brown skin and give it a patina of sickness. Illuminated from above, the fishman's forehead and nose gleam as if covered in slime; his lips are all too visceral, reminiscent of exposed organs or uncooked meat. The lips are also the creature's most hybrid feature: recognizably human closer to the mouth, and fish-like at the outer edges. The fishman's glazed look is directed outward to the very future that WWF anticipates and the creature already inhabits. In its rheumy eyes and shrunken, but recognizably human, ears, the viewer sees both the terrifying consequences of change and a failure to perceive the horrors that were awaiting him when he still had time to act.

Just as we have entered the Anthropocene, a new epoch in which human behaviors modify geophysical systems, we have transformed the matter of anthropos.[1] Fearing that the damages of climate change are rapidly becoming the new normal, WWF assigns to "change" its discomfiting meanings and confers on people agency to stop deleterious transformations that result from human actions. The ad reminds viewers that human bodies are vulnerable to environmental pressures, that human genes register the weirded world we create. But the ad is not without hope, that other buzzword of the Obama 2008 campaign. If humans have the enormous power to reverse the timeline of evolution, might we not heed the urgent tagline that puts power in our hands and address the conditions that lead to mutation?

WWF's "Stop Climate Change" campaign takes seriously Lawrence Buell's assertion that "environmental rhetoric rightfully rests on moral and especially aesthetic grounds *rather than* scientific [grounds]."[2] The organization builds its message using a visual rhetoric that mobilizes the disgusting, a powerfully moral affect, as we'll soon see. Twenty-first-century environmental activism now requires the conventions of horror and the negative aesthetics of the disturbing and the disconcerting. The activists of the UK group Plane Stupid evidence this trend. Through "subtervising" and public action campaigns, the British environmentalists hope to "bring the aviation industry back down to earth!" as their motto announces.[3] Its 2009 "Polar Bears" video, developed by the London-based Mother creative agency, played before films and then went viral as environmental bloggers debated the ad's controversial approach. The thirty-second film shows polar bears careering off of skyscrapers and splatting onto the sidewalks and cars parked below (figures 4.2 and 4.3).[4] It ends with the information that the average flight within Europe emits greenhouse gases equal in weight to an adult polar bear. The scene looks postapocalyptic; though set in a city, it is devoid of people and captures only barren trees and lifeless streets. The sole sounds are of a jet flying overhead and of the bears' bodies slapping against concrete and metal. Whereas the fishman is fantastic, though made plausible

FIGURE 4.1. "Stop Climate Change Before It Changes You," 2008. Germaine & Co for WWF. Photographer, Christophe Gilbert, www.christophegilbert.com. *Printed with permission of Christophe Gilbert and Germaine & Co.*

through realist details, the bears are utterly lifelike. Environmental campaigns usually trot out polar bears to raise awareness of global warming and shrinking polar icecaps and present them as cuddly infants or majestic, maternal adults. Plane Stupid instead mangles them and splashes them with blood.

Undoubtedly, environmentalists still rely on familiar rhetorics of nostalgia for a lost Eden, of anticipatory guilt, or, as the previous chapter attests, of eye-opening wonder to activate environmental involvement, but WWF and Plane Stupid are joining artists in rethinking how they communicate the prospect of an endangered future. And, for these groups as for the authors of ecosickness, the violable boundaries of bodies excite creative energy. Environmental activists are catching on to what contemporary writers have known for the past several decades: that, in Timothy Morton's words, "ecological art is duty bound to hold the slimy in view."[5] Rather than activate the cuddle response with cute photos of frolicking animals, WWF imagines humans becoming animal and Plane Stupid brings animals' insides out, smearing them onto cinema screens. They address environmental catastrophe by eliciting disgust through monstrous hybridity and gory death. The WWF ad concocts a powerful admixture of animality and vulnerability to arouse disgust. The fishman reminds us that, while human, we are at base animals. It depicts "violations of the exterior envelope of the body," which psychologists Paul Rozin, Jonathan Haidt, and Clark R. McCauley show to be strong elicitors of disgust.[6] In "Polar Bears" the animal's "envelope" ruptures, and viewers confront the death of individual creatures and, by implication, of an entire species. Crucially, though, the moral charge of disgust is directed not at the figures in the image but at the unspoken human subject of the imperative copy in the WWF ad. Sparking the "horrified curiosity" that correlates to disgust for activist ends, these campaigns join environmentally focused fiction in capitalizing on what David Pole calls the "striking" feature of disgusting objects: "that they always seem to matter, to fix attention."[7] Disgust slaps us in the face and forces us to confront that which we would rather ignore.

Arguably, disgust's unignorable quality and its association with representational excess made it appealing to David Foster Wallace when he wrote *Infinite Jest* (1996), a narrative about how crippling

FIGURES 4.2 and 4.3. Stills from "Polar Bears," 2009. Mother London for Plane Stupid. *Printed with permission of Joe Ryle for Plane Stupid*

forms of attention have shut out more nourishing varieties. When compared to the WWF and Plane Stupid projects, Wallace's novel is less obviously mobilizing disgust in the service of an environmental message. However, this chapter will show that disgust in *Infinite Jest* combats what Wallace describes to Larry McCaffery as a "contemporary condition [that] is hopelessly shitty, insipid, materialistic, emotionally retarded, sadomasochistic and stupid" and that the shittiness of this condition depends on environmental factors.[8] Handed this situation, the novelist must cultivate readers' "capacity for joy, charity, genuine connections" by "'author[ing] things that both restructure worlds and make living people feel stuff.'"[9] The novel, then, is not only an imaginary world; it can reconfigure the world beyond its pages by modeling and generating emotion, emotion that might not be so pleasant. Wallace's project raises an important question for *Ecosickness in Contemporary U.S. Fiction*: how can aesthetic forms produce affects that enhance an audience's awareness of "hopeless" social and material conditions? In particular, how does a novel transform disgust, which takes us "fromward something," into an affect that serves the restructuring ends that Wallace assigns to fiction?[10] To understand how disgust can be aversive but can also motivate ethical relations, this chapter brings *Infinite Jest*'s underexamined environmental and medical imaginaries into the interpretive frame.

Composed beginning in 1991, *Infinite Jest* speculates that the first decade of the next millennium will exhibit the ills that Wallace diagnoses in his 1993 interview with McCaffery: gross materialism, isolation, and emotional vapidity. Both time and space have been radically reconfigured in Wallace's storyworld. Corporations begin sponsoring time in 2002, the Year of the Whopper, to recover advertising revenue lost with the demise of broadcast television.[11] By the Year of Glad (2009), when the novel opens, the United States, Mexico, and Canada have merged to form the Organization of North American Nations, or O.N.A.N.[12] Because the U.S. is choking on the effluvia of its hyperconsuming society, it forces its impotent northern neighbor to annex portions of northern New England that have been turned into a massive waste dump. From the reshaping of the continent, the novel zooms in on the reshaping of smaller spaces where its dominant plots take place: the Boston conurbation, Phoenix, and Tucson's

Tortolita foothills. In the fictional Enfield neighborhood of Boston, the narration moves between two institutions whose geographical locations symbolize their missions. The Enfield Tennis Academy (E.T.A.) occupies a site engineered by the school's founder, Jim Incandenza, to attract "boys [who] like great perspectival heights and spectacular views encompassing huge swaths of territory."[13] By "balding and shaving flat the top of the big abrupt hill," Jim creates a setting that offers vistas of Boston's diverse topography: "Commonwealth Avenue's acclivated migration out of the squalor of Lower Brighton," "the spiky elegance of B[oston] C[ollege] and the broad gentrification of Newton," and the "transformers and high-voltage grids and coaxial chokers" of a nearby power plant (241–42). Literally above it all, E.T.A. is a site for lofty ratiocination and abstraction as well as for physical conditioning that produces grotesquely fit bodies. By contrast, Ennet House Drug and Alcohol Recovery House sits in the shadow of Incandenza's hill and is a refuge for those who have hit a figurative "Bottom" (347). The halfway house welcomes illogic; residents are urged to remain grounded by casting off thought because the head is the enemy: "most Substance-addicted people are also addicted to thinking" (203). The dilapidated house's open plan discourages the secrecy associated with drug use and facilitates interaction. The doors have no locks so that people and feeling might flow unimpeded.

These spatial arrangements express and enforce many of the ethical and social concerns that give *Infinite Jest* its thematic heft. Moreover, this quick overview of the novel's key settings points to one of this chapter's claims: if novels are the empathy engines that Wallace wants them to be—if, in other words, they provide "imaginative access to other selves"[14]—the texts' environments are crucial components. In Wallace's storyworld, environmental injustices arise from the affective and spatial detachment that contemporary media culture and politics promote. This chapter begins by delineating the forms of detachment that attract Wallace's satirical eye, motivate the novel's main plots, and distinguish its style. Detachment is not only an epiphenomenon of psychological and social decay; it is also, and crucially, an environmental problem that human bodies register. These claims set a foundation for analyzing *Infinite Jest* as an ecosickness fiction that articulates the conceptual and material breakdown of the body-environment boundary and establishes the body as a primary conduit to environmental understanding. Wallace's fiction of social, ecological, and somatic poisoning entangles body and environment causally and conceptually. *Infinite Jest* thereby molds a medicalized environmental consciousness that, like the WWF and Plane Stupid ads, entails the affect of disgust. After first considering whether simply getting out into the world is a path toward connection, I examine disgust as an unlikely affect of attachment that arises as the narrative imbricates body and environment. Read with accounts of disgust from across cultural theory, philosophy, and cognitive science, *Infinite Jest*'s scenes of disgust establish how this emotion can goad a person to social and environmental investment.

Ultimately, I argue for disgust's transformative role in the canon of environmental emotions, a role that WWF and Plane Stupid exploit in their campaigns a decade after the publication of *Infinite Jest*. Unlike these groups, however, Wallace's fiction dramatizes sick, permeable bodies and permeable, medicalized spaces to awaken readers to the idea that we are affective beings gathered into networks of obligation with other beings. It does so by mobilizing the defamiliarizing affect of disgust. Unlike the traditionally positive emotion of wonder that, in Powers's work, creates problems of care through complex connectivity, the unpleasant emotion of disgust paradoxically promotes connection and ethical concern. *Infinite Jest* thus depicts ecosickness through an ugly aesthetic that points a way out of the sicknesses endemic to contemporary reality.

DETACHED DISPOSITIONS

One complaint houses all the problems that mar *Infinite Jest*'s fictionalized U.S.: people and the cultural artifacts they produce are utterly self-involved in their attempts to escape the pains of life. They, as Wallace writes in "E Unibus Pluram: Television and U.S. Fiction," "receive without giving."[15] Solipsism and self-indulgence, states in which the individual is the measure of and guide for all experience, are psychological analogues for the uncritical self-reflexive style of postmodern culture. What troubles Wallace troubles like-minded critics of postmodernism: in retreating into the self or reveling in self-referential artistic forms, a person risks distancing herself from the world to the point where she utterly detaches herself from social relations. A psychological disposition with spatial and political dimensions, detachment is the prime mover of *Infinite Jest*'s plots about an addictive film and insurgency against new forms of U.S. imperialism. To condemn detachment as a limit on intersubjective relations, the novel uses grammatical forms that distance—notably, passive voice—and presents the logic of meta-emotions, feelings about emotion that remove an individual from the primary pain. Examined together, the thematic, grammatical, and affective expressions of detachment conduct readers from the personal ramifications of pervasive, disabling alienation to its geopolitical and environmental ramifications.

Harold (Hal) Incandenza is the characterological center for the novel's critique of detachment. Hal is the youngest of James (Jim) and Avril Incandenza's three sons and the protagonist of the plots based at E.T.A. Intellectually and athletically gifted, Hal is impassive to a fault and nurses an addiction not so much to marijuana as to the rituals that accompany his indulgence in it. Hiding his drug use feeds Hal's habit of emotionally distancing himself from others; it also guarantees that he can still slam opponents on the court and excel at his studies. The Incandenzas pick up on Hal's coldness in different ways. The patriarch sees a teen who has retreated inward to the point where he no longer communicates with his family. The novel suggests that the actual problem is that Jim simply can no longer hear his son, but the father still launches an attack against the teen's retreat into himself. Before he commits suicide during alcohol withdrawal, Jim produces a film titled "Infinite Jest" that he intends to counteract Hal's solipsism, to warm up his emotional core and encourage interpersonal connection. Despite—or because of—its experimental techniques, the film lives up to its nickname, "the Entertainment," in a disastrous way.[16] It is so addictively pleasurable that it induces a state of catatonia that eventually leads to the viewer's death from self-neglect. Due to the film's suicidal effects, various parties seek it for political ends. A group of Québécois separatists want to use it as a cultural weapon: after rendering the U.S. impotent, they will secede from both O.N.A.N. and Canada. The U.S. "Office of Unspecified Services," the intelligence arm of the federal government, seeks it to thwart the separatists' plots.

The politically motivated hunt for "Infinite Jest" unfolds over many pages, but readers most fully grasp the connection between Jim's masterpiece and detachment at the novel's end, when Jim's ghost haunts the comatose Don Gately. Gately is the center of the Ennet House narrative. A recovering addict and supervisor at the residence, he sustains a life-threatening gunshot wound when he intervenes in a violent dispute between another resident, Randy Lenz, and the owners of a dog Lenz killed (608–15). Jim explains to Gately that he had stood witness as Hal became "a steadily more and more *hidden* boy . . . and no one else in the wraith and the boy's nuclear family would see or acknowledge this, the fact that the graceful and marvelous boy was disappearing right before their eyes. They looked but did not see his invisibility" (838). Jim's admittedly oblique countermeasure to Hal's transformation was to

> spen[d] the whole sober last ninety days of his animate life working tirelessly to contrive . . . something the boy would love enough to induce him to open his mouth and come out—even if it was only to ask for more. . . . His last resort: entertainment. Make something so bloody compelling it would reverse thrust on a young self's fall into the womb of solipsism, anhedonia, death in life. . . . To bring him "out of himself," as they say. The womb could be used both ways. A way to say I AM SO VERY, VERY SORRY and have it *heard*. (838–39)

Art must counteract a detachment that amounts to living death. Jim films his muse Joelle van Dyne, a future Ennet House resident, leaning over a crib and repeating the phrase, "I'm so sorry. I'm so terribly sorry," for twenty minutes. He shoots Joelle from the perspective of the child, using a camera with an "ocular wobble" to replicate a neonate's blurred vision (938–39).[17] Even though the film suggests that giving life is an act that demands penance, it also implies that depicting birth is still the best foil to death.

Jim's fight against solipsism, though targeting only one person, echoes Wallace's program for a new fiction. Rather than bury the reader in self-reference and metafiction for its own sake, the responsible novelist should ventriloquize and broadcast others' voices. "We all suffer alone in the real world," Wallace explains with urgent pathos;

"true empathy's impossible. But if a piece of fiction can allow us imaginatively to identify with characters' pain, we might then also more easily conceive of others identifying with our own. This is nourishing, redemptive; we become less alone inside."[18] Wallace's comments to McCaffery appeared in the *Review of Contemporary Fiction* and therefore reached an audience of peer novelists, literary readers, and critics. The interview transcript preceded "E Unibus Pluram," his essay on the state of the American novel, in which Wallace passionately and despairingly critiques contemporary novelists who embrace an irony whose ends have been distorted thanks to television's appropriation of it.[19] In the 1960s and 1970s leading postmodern novelists such as Thomas Pynchon and Ken Kesey employed irony with idealistic intentions. They assumed "that etiology and diagnosis pointed toward cure, that a revelation of imprisonment led to freedom" from Americans' subservience to corporate and governmental hypocrisy (66–67). In the decade from which Wallace writes, irony has mutated. The ironic detachment that pervades what he calls "Image-fiction" "is both medicine and poison": it gestures toward a cure, but only exacerbates the solipsism that the content and form of television already encourage (25–26).[20] Late twentieth-century writers must rethink, not recirculate, the stultifying irony that surrounds them because irony is the aesthetic corollary to solipsism and cynicism; it blocks the capacity for feeling that's at the root of interpersonal connection. As Wallace writes in "Joseph Frank's Dostoevsky," the point of life, as of literature, is not "simply to undergo [or represent] as little pain and as much pleasure as possible;" avoiding pain only ends up being "awful lonely."[21] Writers must own up to their influence and halt the unredemptive "cat's-away-let's-play Dionysian revel."[22] The medical diction—"etiology and diagnosis," "cure," "medicine and poison"— that permeates "E Unibus Pluram" sharply hones the novelist's job description: she can, indeed *must*, be a healer.

Held up to this measure, Jim Incandenza's efforts fail.[23] The film never reaches Hal to halt his slip into "death in life" and instead misfires, proliferating addicts who, enthralled by the work's pleasures, end up "in exile from reality" (20). The film is all too potent, "so bloody compelling" that it drives viewers to "ask for more" at the expense of social relationships and even survival (839). For different reasons and with deliberation, Wallace's *Infinite Jest* also reproduces that which it critiques. Focusing its eagle-eyed satire on the pervasive social detachment in the contemporary U.S., the novel inscribes detachment into the grammatical form and emotional logic of the narrative.[24]

Passive voice, which abounds in *Infinite Jest*'s involuted sentences, speaks detachment because it separates an actor from the action that she performs. The description of Joelle van Dyne's overdose provides one example of how the passive inscribes detachment at the level of the sentence. A thirty-seven-line sentence narrates Joelle's suicide attempt. She enters her friend Molly Notkin's bathroom and commences a ritual of freebasing cocaine that she hopes will kill her, but that only lands her in Ennet House. The marathon sentence takes in the bathroom's design, the cacophony of the party outside, and the effects of the coke's alluring blue smoke. As a sign that she has reached a point where her high is "so good she can't stand it," "Joelle's limbs have been removed to a distance where their acknowledgement of her commands seems like magic, both clogs simply gone, nowhere in sight, and socks oddly wet" (240). The cocaine has not only figuratively amputated her limbs, that is, "removed [them] to a distance;" it has detached her from her self-command. Her arms move from offstage, as if by "magic." Drug use exemplifies the detached position of citizens of the present, and the passive voice here communicates one way in which this habit hobbles: it detaches a person from her own body, takes her out of her skin, and cuts the connection between brain and appendage. In this key passage, *Infinite Jest*'s form fuses with one of its dominant tropes. Joelle has removed herself from the revels and sequestered herself in the bathroom to freebase, and this act experientially removes her from herself. The passive conveys the material detachment in which drug use, among other isolating behaviors, ends.

This pattern of detachment permeates *Infinite Jest*'s representations of characters' affect as well. The novel's psychologically embattled cast only feels emotion as if removed from it. That is to say that "meta-responses," or emotions about emotions, supplant characters' primary responses. Metaresponses are reactions not to immediate "eliciting conditions"[25]—for example, feeling envious of a neighbor's wealth—but to "how one feels about and what one thinks about one's responding (directly) in the way one does"[26]—feeling ashamed *of* that envy. Across the spectrum of affective content that *Infinite Jest* depicts, from the pleasures of smoking a joint to the agonies of wrenching depression, the novel theorizes the emotional bearing of twenty-first-century subjects as detached in this experiential sense.

Geoffrey Day's account of excruciating depression illuminates this relation. "'[Depression] was a bit like a sail, or a small part of the wing of something far too large to be seen in totality,'" he describes. "'It was total psychic horror: death, decay, dissolution, cold empty black malevolent lonely voided space.' . . . 'I understood what people meant by *hell*. They did not mean the black sail. They meant the associated feelings'" (650–51). An expanse opens up between the character and the primary pain of depression; into this expanse, second-order "associated feelings" rush in. The onslaught of meta-affect torments Geoffrey, even as it distances him from the sources of the depression itself. Solidifying this metalogic, the form of the sentences about metaresponses carries detachment into the reading experience. Here and throughout his writing, Wallace strings together possessive prepositional phrases—"a small part *of* the wing *of* something far too large"—with the effect that the reader's attention moves away from "something" itself to a synecdoche for it. The metalogic of emotions appears elsewhere in Wallace's stories from the late 1990s; they too depict the auxiliary affects that overload a subject and distance her from primary emotion. In "The Depressed Person" (1998), the titular protagonist recalls her overpriced therapy sessions and decides that she is a failed patient because "she felt able to share only painful circumstances and historical insights about her depression and its etiology and texture and numerous symptoms instead of feeling truly able to communicate and articulate and express the depression's terrible unceasing agony *itself*."[27] The accessories to anguish preoccupy the woman and sap her emotional energies. They insert a gap between feeler and primary feeling, and her depression intensifies into isolation from self as well as world.

Grammatical form and strategies for depicting emotion thread detachment through the stylistic and experiential fabric of *Infinite Jest*. These strategies amplify the sweeping critique of solipsism that unfolds through plot and character development. Condemning detachment, the novel revises understanding of the critical potential of distance. In his seminal theorization of postmodernism, Fredric Jameson complains that "distance in general (including 'critical distance' in particular) has very precisely been abolished in the new space of postmodernism. We are submerged in its henceforth filled and suffused volumes to the point where our now postmodern bodies are bereft of spatial coordinates and practically (let alone theoretically) incapable of distantiation."[28] Jameson analogizes spatial and theoretical conceptions of "distance." In other words, he equates the kind of detachment that enables critique to physical distances that put space between objects. For Jameson, the cultural realm is no longer even partially autonomous, and we, in turn, are suffocating on a culture in the thrall of late capitalism. By contrast, *Infinite Jest* posits that distance has gone too far. While distance might have been a valuable instrument in the critical toolbox in the modern and even early postmodern periods, it is now anaesthetizing rather than productive.[29]

The comparison to Jameson helps show detachment to be an intersubjective and psychological disposition that ramifies politically and ethically. Detachment accrues eco- and geopolitical significance in *Infinite Jest* as the novel makes detachment the psychological underpinning of environmental reconfiguration. The pervasiveness of distancing in the novel thus suggests that, just as affective detachment can compromise sociality, affective attachment might counterbalance injustice. In order to imagine attachment, the novel attunes readers to the literal and figurative spaces that must be traversed in pursuit of connection.

"EXPERIAL" AMBITIONS

O.N.A.N. is the apotheosis of the "Cornucopia City" that Vance Packard invented for *The Waste Makers* (1960). Packard's book elaborates how a U.S. economy driven by consumption and planned obsolescence produces physical waste and psychic discontentment. It narrates this condition through an allegory that resonates with Wallace's fiction. O.N.A.N., like Cornucopia City, runs on a "hyperthyroid economy that can be sustained only by constant stimulation of the people and their leaders to be more prodigal."[30] Americans accept this agenda in the misguided hope that constantly acquiring new things will enliven prosaic existence and will distinguish them from the homogeneous crowd.[31] Unbridled consumption misses these targets, however, and instead "'deadens sensitivity to other human beings.'"[32] In *Infinite Jest*, prodigal consumption leads to prodigal waste, and the government's moves to manage the nation's trash lead to geopolitical schemes that reveal its "deadened sensitivity." The reigning U.S. administration establishes a putatively collaborative alliance with its neighbors, yet this political relationship is just as self-serving as the personal ones that distress Wallace. Under "Interdependence," the U.S. strong-arms Canada into accepting land on the border for a vast toxic waste dump and energy production site. The policy of "Interdependence" is, as Katherine Hayles aptly glosses, "merely rampant nationalism under another guise," an assertion of absolute American sovereignty.[33] "The unpleasant debris of a throw-away past" is asphyxiating America; conducting that debris across another nation's borders, the U.S. can enlarge its passageways to breathe easier (382). The country thus enters the business of exporting waste, of "sending from yourself what you hope will not return" (1031*n*168). Needing to put distance between itself and its waste, a source of opprobrium and fear, the nation rigs international relations and space itself. Under the organizing concept of detachment, then, *Infinite Jest*'s psychological and social climates hook up with its ecopolitical arrangements.[34]

The U.S. adopts its waste export policy at a time when "all landfills got full and all grapes were raisins and sometimes in some places the falling rain clunked instead of splatted" (382). With the obsessive-compulsive Johnny Gentle at its head, the Clean U.S. Party (C.U.S.P.) capitalizes on environmental decline and rises to power under the motto "Let's Shoot Our Wastes into Space" for a "Tighter, Tidier Nation" (382). Closer than outer space and just as politically impotent, Canada is the ideal destination for the nation's trash. Merging with Canada on paper but not in spirit, C.U.S.P. inaugurates a new geopolitical and environmental regime: rather than pillage other nations' resources to meet its own industrial demands, as in *im*perialism, the U.S. banishes the by-products of capitalism under *ex*perialism.[35]

For this waste export plan to succeed, the American government manipulates intracontinental relations and territory through a program of environmental reconfiguration and semantic obfuscation. Appealing to Canada's cooperative spirit, Gentle's cabinet relocates residents of the northeastern U.S. and southeastern Canada, an area now known as "the Great Concavity" (or "Convexity" from north of the border), so that it can catapult its unwanted refuse into the nearly vacated territory.

Through a puppet show created by the second Incandenza son, Mario, readers learn the history of O.N.A.N.'s inception and the environmental injustices that the U.S. perpetuates under its aegis.[36] Every Interdependence Day (November 8), Mario's characters reenact the cabinet meetings in which the U.S. government reshaped the continent:

> Tine [future head of O.N.A.N. intelligence] places two large maps . . . on Govt.-issue easels. They look both to be of the good old U.S.A. . . . The second North American map looks neither old nor all that good, traditionally speaking. It has a concavity. It looks sort of like some person or persons have taken a deep wicked canine-intensive bite out of its upper right bit, in which an ascending and then descending line has its near-right-angle at what looks to be the historic and now hideously befouled Ticonderoga NY . . .
> Sec. State: A kind of ecological gerrymandering?
> Tine: The president invites you gentlemen to conceive these two visuals as a sort of before-and-after representation of "projected intra-O.N.A.N. territorial reallocations," or some public term like that. *Redemisement*'s probably too technical. . . .
> Sec. State: Still don't see why not just retain cartographic title to the toxified areas, relocate citizenry and portable capital, use them as our designated disposal area. Sort of the back of the hall closet or special wastebasket underneath the national kitchen sink as it were. (403–5)

Through "ecological gerrymandering," C.U.S.P.'s antiwaste platform becomes foreign policy. More than just gerrymander the territory—

that is, reconfigure political boundaries that are always virtual—the administration pollutes and thereby materially transforms the very spaces that those boundaries delimit. On the map proposing this policy, the Concavity is symbolically detached from its home nations: it has been bitten off. Ecological gerrymandering is thus an environmental expression of the distancing and detachment that degrades social relations. By quarantining contamination, by giving it a "special wastebasket underneath the national kitchen sink," but then sliding that wastebasket across an extant national border, the U.S. hopes to neutralize the "Menace" of waste and distance itself from the ugly, transnational consequences of producing, consuming, and disposing stuff (382). Gentle's proposal and the opaque "public term[s]" that communicate it to the people are designed to obscure the fact that displacing waste requires displacing people and altering ecosystems. Indeed, as this chapter makes clear, displacement is just one of the social and environmental injustices on which American territorial, economic, and energy security rest in *Infinite Jest*.

Environmental conditions are integral to Wallace's imagination of new social arrangements in his previous works as in the 1996 novel. Indeed, the fictional manipulation of land represents America's advance toward a society defined by its collective isolated individualism, a society to which the motto "e unibus pluram" ("from one, many") applies. *The Broom of the System* (1987), Wallace's debut novel, first entertains how manufacturing space not only expresses an ideology but also creates one. In a scene providing the backstory for the diegetic present, the governor of Ohio and his cronies develop a plan to rekindle Americans' intrepid spirit to stave off the complacency that suburbanization and consumerism have induced. Their solution: the Great Ohio Desert, or G.O.D. "'We need a wasteland,'" Governor Zusatz decrees. "'A desert. A point of savage reference for the good people of Ohio. A place to fear and love. A blasted region. Something to remind us of what we hewed out of.'"[37] To counteract the developed wasteland of strip malls, gas stations, and fast food chains, Zusatz will construct a wilderness *qua* wasteland through which Ohioans can build their mettle. Engineering environments, the governor will reinstall an ideology of rugged individualism and self-reliance. In a similar vein, C.U.S.P. takes advantage of Americans' need for an enemy: it promotes pollution to the status of necessary national terror for a "post-Soviet and -Jihad era" and then proposes extreme solutions for displacing and eradicating the threat (382).

In *Infinite Jest* an individual pathology guides contemporary environmental policy rather than the national myth that inspires Zusatz. President Johnny Gentle is "a world-class retentive, the late-Howard-Hughes kind, the really severe kind, the kind with the paralyzing fear of free-floating contamination," and he successfully transmits his compulsive aversion to waste to the nation that he rules (381). The president's skewed environmentalism contrasts more familiar forms of ecoconsciousness associated with the deep ecology and back-to-nature movements. Despite their different philosophical bases, both programs advocate renouncing consumption and turning back the clock on modernization. Under Gentle, *aestheticism* rather than *asceticism* becomes the point of advocacy for a green-washed political movement in the twenty-first century. This shift is not secret. As the narrator explains, "C.U.S.P. had been totally up-front about seeing American renewal as an essentially aesthetic affair" (383). Through C.U.S.P.'s ambitions, the novel satirizes the Keep America Beautiful organization that formed in 1953 with a mission of "bringing the public and private sectors together to develop and promote a national cleanliness ethic."[38] *Infinite Jest* speculates that Keep America Beautiful could become a full-blown political party. The novel thus imagines an aesthetic stance generating a perverse environmental politics of "ecological gerrymandering" that depends on the same detachment that compromises psychic and ethical life in the contemporary U.S. The nation's territorial and environmental policies express its "deaden[ed] sensitivity to other human beings."[39] Gentle's personal pathology inspires these policies that alter entire ecosystems, but they ultimately result in the sicknesses of other individuals. While play and humor suffuse these satirical scenes, the linked policies of ecological gerrymandering and waste exportation make clear that environmental despoliation is not simply an "aesthetic affair" in the novel (383). These actions, and the disposition of detachment that they express, have bodily effects. Elaborating these effects, *Infinite Jest* activates an environmental consciousness based in the sick, medicalized body.

For this reason, *Infinite Jest* belongs to the canon of environmental texts that show the human body to be inextricable from the spaces that it inhabits, in literature as in life. If Rachel Carson's *Silent Spring* thrust this idea into the national spotlight in 1962 by warning that human and nonhuman animals carry the burden of chemical contamination, Wallace's novel gives literary focus to her message of pervasive toxicity and the body's vulnerability to it. Historian Linda Nash demonstrates that "an ecological view of the body" that conceived of humans as enmeshed in their environments was not new when Carson wrote about human ecology.[40] However, the idea of the ecological body "characterized by a constant exchange between inside and outside, by fluxes and flows, and by its close dependence on the surrounding environment" fell by the wayside.[41] Corporate obfuscation and biomedical protocols have ensured that notions of the "modern body," that is, the body "composed of discrete parts and bounded by its skin," still dominates fifty years after *Silent Spring*.[42] By amplifying the effects of toxicity on human characters and strange environments, *Infinite Jest* revives the ecological body. In this respect, it participates in the cultural production of "transcorporeality" that Stacy Alaimo details in *Bodily Natures*. Wallace's text too narrates "the material interconnections of human corporeality with the more-than-human world" that "often revea[l] global networks of social injustice . . . and environmental degradation."[43] And, as in the novels, films, memoirs, and activist campaigns that concern Alaimo, *Infinite Jest* depicts these interconnections in the technical languages of science: medicine, psychology, and pharmacology, most saliently. The novel's environmental consciousness is thus thoroughly medicalized.[44] Obsessive compulsion, a medical disorder, leads to a dysfunctional, detached relation to the environment and to human others, and this detachment underpins the reshaping of space, geopolitics, and, as I establish here, human cells, tissues, and organs.

Katherine Hayles first directed critical attention to *Infinite Jest*'s ecological imagination in arguing that the novel "recogniz[es] that market and individual, civilization and wilderness, coproduce each other."[45] Ultimately, she detects in the novel an ecological perspective through which we "discover the text's recursive patterns so we can see it, as well as the world it describes, as a complex system that binds us into its interconnections, thus puncturing the illusion of autonomous selfhood."[46] Hayles identifies the overlapping spheres of awareness that *Infinite Jest* aggregates, and she moves discussion of the novel outside of established interpretive frames of postmodernist irony and encyclopedism. However, she does not fully elucidate how "real ecologies" figure in the novel's scheme of cultivating connection against the "illusion of autonomy."

After C.U.S.P.'s ecological gerrymandering, two distinct environments evolve: the toxic Great Concavity and everything south of it. Conceiving the former, Wallace flourishes his talent for bleak humor. A spirited mythology builds up around the toxic zone that is off-limits to civilians. As E.T.A. students roam the campus tunnels, they terrorize each other with tales of the mutant rodents that have migrated from the Concavity to the school's trash-strewn subterranean zone. The narrator checks the students' wild ideas, assuring readers that "feral hamsters . . . are rarely sighted south of the Lucite walls and ATHSCME'd checkpoints that delimit the Great Concavity" (670).[47] But no one doubts that these creatures populate Canada, and their terrifying aspect takes hold of the public imagination. "Bogey-wise [they're] right up there with mile-high toddlers, skull-deprived wraiths, carnivorous flora, and marsh-gas that melts your face off and leaves you with exposed gray-and-red facial musculature for the rest of your ghoulish-pariah life" (670). Hal rehearses the most common tale about the region's mutant creatures when he expounds why Québec is a hotbed for extremism:

> "It's mostly now western Québecer [sic] kids the size of Volkswagens shlumpfing around with no skulls. It's Québecers with cloracne [sic] and tremors and olfactory hallucinations and infants born with just one eye in the middle of their forehead. It's eastern Québec that gets green sunsets and indigo rivers and grotesquely asymmetrical snow-crystals and front lawns they have to beat back with a machete to get to their driveways. They get the feral-hamster incursions and the Infant-depredations and the corrosive fogs. . . . Proportionally speaking it's Québec that's borne the brunt of what Canada had to take." (1017*n*110)

The passage's tone is playful, but the allusion to chloracne, a breakout of cysts resulting from dioxin exposure, introduces something more horrific into this humorous exposition. The irradiated rats and infants that scurry about the Concavity draw on the vocabulary of nuclear horror from the cold war but, like WWF's "fishman" campaign, *Infinite Jest* uses disgust to trigger concern about the coupled contamination of bodies and ecosystems. In contrast to a text like *Silent Spring*, whose title introduces the imagery of absence and extinction that will illustrate its argument, Wallace's novel imagines toxification through both diminishment and augmentation. As this passage already hints with the mention of one-eyed infants, diminished function counterweighs abundance in the lethal equation that drives U.S. environmental policy. The grotesque physical extremes of diminution and gigantism are analogues for the extreme environments that result from America's misguided response to consumption and waste.

Sending waste into the Concavity is only one way that C.U.S.P. exploits the O.N.A.N. alliance. The Concavity also has "to take" the effects of "annular fusion," a form of energy production invented by Jim Incandenza in which excess breeds excess ad infinitum. In brief, it is "'a type of fusion that can produce waste that's fuel for a process whose waste is fuel for the fusion'" (572). A complex closed-loop process—in which waste produces energy that produces waste to fuel the cycle all over again—generates an equally complex ecological cycle in the affected region. During the fusion process, all toxins are sucked out of the ecosystem, and this arrests "organic growth for hundreds of radial clicks in every direction" to produce a barren wasteland (573). When the toxins are pumped back in, the legendarily overabundant environment that Hal describes flourishes. We do not need to grasp all the details of this involuted cycle in order to see the impacts of generating energy from waste. As the U.S. pursues energy independence, poisons produce poisons without end. Consequently, the surrounding environment vacillates between the extremes of a lifeless desert and a verdant rain forest. Imagining territorial reconfiguration, toxification, and energy production, Wallace presents a predicament that is acutely familiar to twenty-first-century Americans: consumer and industrial growth fuels the demand for energy, the production of which alters the environment irremediably.[48] Importantly for my analysis of the affects of ecosickness in the novel, the effects of waste displacement and annular fusion are not limited to aberrant sunsets, snow-crystals, and lawns. Ultimately, the human body is the point of application for the unbridled toxification that results from consumerism, industrialism, and social and environmental detachment.

Infinite Jest spins the somatic consequences of experialism and O.N.A.N.'s environmental policies through Rémy Marathe, the paraplegic leader of a group of Québécois separatists, "Les Assassins des Fauteuils Rollents [*sic*]" (A.F.R.). Emphasizing the materiality of environmental poisoning, Marathe catalogs his wife Gertraude's contaminated body to convince Hugh ("Helen") Steeply, his collaborator in U.S. intelligence, that O.N.A.N. policy is medically unjust. In English marked with French grammatical tics, he enumerates Gertraude's deformities:

> "She had no skull, this woman. Later I am learning she had been among the first Swiss children of southwestern Switzerland to become born without a skull, from the toxicities in association of our enemy's invasion.... Without the confinement of the metal hat the head hung from the shoulders like the half-filled balloon or empty bag . . . Her head it had also neither muscles nor nerves. . . . There was the trouble of the digestive tracking. There were seizures also. There were progressive decays of circulation and vessel, which calls itself restenosis." (778–79)

The medical blazon spans several more pages. It is worth noting that the pace and intensity of Wallace's proliferative style do not let up when the narration turns to descriptions of lack and deficiency, signaled by "half-filled," "empty," and "restenosis" (that is, the narrowing of blood vessels). As Americans distance themselves from their wasteful habits, they also jettison from view the ethical implications of experialism and ecological gerrymandering. The A.F.R.'s insurgency focuses attention squarely on the harms of O.N.A.N. environmental and energy politics. Gertraude's ruined body and health fuel Marathe's rebellion against the U.S. and turn him into a double (or quadruple [94]) agent who crosses the A.F.R. to get medical assistance for his wife. Like the waste that's fated to return, the nation's toxic policies come back to haunt it in the form of damaged bodies that ignite political hostilities. In this way, the entanglement of environmental change and somatic sickness motivates the separatist plot involving Marathe. *Infinite Jest*'s almost Rabelaisian descriptions of poisoned bodies highlight not only that the toxification of space expresses itself through bodies but also that the body is the way that we come to understand environmental manipulation as an injurious practice on an global scale.

Reconfiguring space, the novel demonstrates that the damages of detachment ramify beyond the interpersonal and pose environmental, somatic, and political threats in the contemporary U.S. C.U.S.P. devises experialism as a way to cover up Americans' profligate habits of producing and disposing, but this program creates environmental and bodily injuries that instead materialize those habits in grotesque form. Ultimately, the text weaves human bodies into their environments in order to make visible and palpable the multifarious damages of detachment. In this way, *Infinite Jest* fleshes out the cultural form that Buell dubs "toxic discourse." Buell defines this mode as "expressed anxiety arising from perceived threat of environmental hazard due to chemical modification by human agency."[49] He historicizes toxic discourse, continuing, "as such, it is by no means unique to the present day, but never before the late twentieth century has it been so vocal, so intense, so pandemic, and so evidentially grounded."[50] While Buell's account of the genres of toxicity is masterfully wide-ranging, he overlooks one of the literary upshots of pervasive toxicity: this ubiquitous "irritant" and the sicknesses that result from it inspire a literary mode in which the medicalized body seeps into environmental consciousness, much as poisons seep into permeable pores.[51] Buell contends that twentieth-century environmental literature cannot help but account for toxicity; I add that late twentieth-century narratives of toxicity cannot help but imbricate mutable human bodies in their imperiled surroundings through the formal and affective strategies to which I turn now.

BODY BUILDING

In *Infinite Jest* environmental manipulation and contamination disrupt ecologies and produce sick bodies that materialize the damages of detachment. To produce an environmental awareness appropriate to the twenty-first century, the narrative causally intertwines body and environment: degradation of the latter disfigures the former. It also elaborates a conceptual relay between these domains in two ways. First, the text expounds its claims for somatic-environmental interdependence by spatializing suffering. That is to say, the narrative presents mental pathologies of detachment like depression using metaphors of space and of geographical orientation. Through its spatial coding of mental sickness, *Infinite Jest* proposes that just being in space is an affliction, but also suggests that we might ameliorate sickness by inhabiting space differently. The second way in which the text conceptually entwines body and environment is by figuring urban environments in terms of the medicalized human body. In particular, it medicalizes space through anatomical conceits: in *Infinite Jest* metro Boston is a body. Wallace joins a lineage of architect-philosophers—from first-century B.C. Roman Vitruvius Pollio to Renaissance man Leon Battista Alberti and his twentieth-century descendent Le Corbusier—who have derived architectural and urban proportions from the human body.[52] In Wallace's late twentieth-century contribution to this tradition, built environments do not only conform to the contours of the body; they also mimic its functions.

The characterization of depression, the most pervasive sickness in *Infinite Jest*, shows the first way in which human and environment are, as Elizabeth Grosz puts it, "cobuilt."[53] Anhedonia is a low-grade form of depression, "a psychological condition characterized by inability to experience pleasure in acts which normally produce it."[54] Anhedonia is the overarching tone of many characters' experience in that being in the world is bereft of pleasure and even emotion unless they are under the influence of drugs or narcotizing entertainments like sitcoms. Much as metaresponses detach individuals from their primary emotions about stimuli, anhedonia detaches people from material reality and other selves. For this reason, anhedonia is an epiphenomenon of social and political arrangements predicated on detachment. An Ennet House resident describes anhedonia and other depression-spectrum disorders in the spatial terms of detachment:

> Kate Gompert's always thought of this anhedonic state as a kind of radical abstracting of everything, a hollowing out of stuff that used to have affective content. . . . The anhedonic can still speak about happiness and meaning et al., but she has become incapable of feeling anything in them, of understanding anything about them . . . Everything becomes an outline of the thing. . . . The world becomes a map of the world. An anhedonic can navigate, but has no location. I.e. the anhedonic becomes, in the lingo of Boston AA, Unable To Identify. (693)

Anhedonics detect "happiness and meaning" but cannot "Identify" with it because they are lifted off the affective terrain peopled by others. Psychic emptiness is akin to having an aerial perspective on the world that provides a vantage on the overall shape of life but evacuates its content. These shapes are devoid of volume; the map has supplanted the world and become pure perspective, incapable of offering "location." In full-on depression, the person emerges from the numbed anhedonic state, but only to confront her own anguish directly. She accesses "an unnumb intuition in which the world is fully rich and animate and un-map-like and also thoroughly painful and malignant and antagonistic to the self" (695). As "The Depressed Person" describes it, one returns to treacherous emotional ground, but is girded in "the psychic armor designed to keep others at a distance so that they (i.e., others) could not get emotionally close enough to the depressed person to inflict any wounds."[55] So while the depressive reinhabits the territory the anhedonic has vacated, she ends up trapped within the private space of her own suffering.

Infinite Jest describes depression in terms that call to mind but also call into question Jameson's demand for improved "cognitive maps" with which to situate ourselves in the late capitalist present. A cognitive map gives us coordinates for making sense of the political, economic, and media structures of capital within which we are enmeshed. It "disalienates" subjects in that it provides them with "a situational representation on the part of the individual subject to that vaster and properly unrepresentable totality which is the ensemble of society's structures as a whole."[56] *Infinite Jest*'s detached subjects will never be able to coordinate the components of contemporary reality into such a map even as—indeed, because—they experience the world *as* a map. In Wallace's imagination the map symbolizes distantiation and disorientation rather than orientation and accessibility. In addition to being alienated from the workings of global capital, characters such as Kate and Hal are alienated from their own subject position. They are "Unable To Identify."

Despite the differences between Jameson and Wallace, they both share a habit of thought. They correlate psychic and material space in order to understand sociocultural and political arrangements in the contemporary U.S. The work of materialist geographers also adds dimension to the dynamic interplay between human psyches and our modified surroundings. In *Uneven Development* (1984), Neil Smith refutes the oft-repeated story that humans exploit, conquer, and reconfigure their surroundings unidirectionally. He urges readers to understand nature as produced rather than as dominated. To this end, he rehearses the Frankfurt School thesis that the spaces in which humans intervene technologically push back by transforming individuals' mental space and affective bearing. *Infinite Jest* offers an instance of this relation. O.N.A.N. manipulates and defiles the environment and only exacerbates the "domination of 'internal nature' (people themselves) and the growing fragility of human existence."[57] In other words, material space and the space of thought and affect are coconstitutive. People develop the means to satisfy their needs, often by dominating other humans and more-than-human nature, and the material outcomes of that process shape human consciousness itself. Edward Soja's cognate arguments extend Smith's analysis in that they dissolve the physical-mental barrier and assert the interdependence of spatial and social forms. Soja contends that "the organization of space [is] not only a social product but simultaneously rebound[s] back to shape social relations."[58] In Wallace's novel the pattern of detachment that manifests in the reconfiguration of material space also shows up in mental dispositions that govern cross-border environmental politics. This does not mean that *Infinite Jest* asserts a causal relationship between O.N.A.N.'s ecological gerrymandering and rampant anhedonia. Rather, by coding endemic mental disorders through spatial metaphors, the narrative correlates material and psychological damage under the sign of detachment.

Equally important to how *Infinite Jest* intertwines earth and soma conceptually is its medicalization of space. Just as Jan Zita Grover medicalizes the Minnesota north woods through descriptions of the trees' "scoriatic bark [and] rheumatoid branches" in *North Enough*, Wallace figures the urban environment in terms of the medicalized human body.[59] Anatomy animates the novel's settings such that contemporary space and the body structure each other. As in Grosz's theory of the corporealized city, body in *Infinite Jest* is "active in the production and transformation of the city. . . . They are mutually defining."[60] She elaborates on this coproduction: "the city is made and made over into the simulacrum of the body, and the body, in its turn, is transformed, 'citified,' urbanized as a distinctively metropolitan body."[61] Just as social norms and values sediment in built spaces and are taken up by the bodies that inhabit and move through them, cities take on the changing forms and functions of the human body. Jameson is again instructive on this matter. Examining the relays between space and bodies under the conditions of late capitalism, he identifies a "mutation in built space itself," a mutation that has outpaced our ability to adapt to it.[62] These new spaces "stan[d] as something like an imperative to grow new organs, to expand our sensorium and our body to some new, yet unimaginable, perhaps ultimately impossible, dimensions."[63] As we see in the lurid descriptions of toxic mutants and disfigured bodies, the mandate to adapt physiologically to spatial "mutation[s]" attains monstrous proportions in *Infinite Jest*. Bodies metamorphose as a result of environmental toxicity, and, crucial to my argument here, environments also mutate into the human body: Wallace grows buildings and landforms into "new organs."[64]

Enfield Tennis Academy sits on the "cyst" of "the whole flexed Enfield limb sleeved in a perimeter layer of light residential and mercantile properties" (241). A growth on Enfield's appendage, E.T.A. is "laid out as a cardioid, with the four main inward-facing bldgs. convexly rounded at the back and sides to yield a cardioid's curve, with the tennis courts and pavilions at the center and the staff and students' parking lots . . . forming the little bashed-in dent that from the air gives the whole facility [a] Valentine-heart aspect" (983*n*3). The branching tunnels that snake under E.T.A. form this heart-shaped institution's veins and arteries, and they channel air to the school's

Pump Room and "Lung," a polyurethane dome that shelters the tennis courts from winter frost.

If E.T.A. is part of Boston's circulatory system, M.I.T.'s Student Union constitutes its nervous system.[65] The Student Union is "one enormous cerebral cortex of reinforced concrete and polymer compounds" (184). There's a lexical shift in this and subsequent passages describing M.I.T.: the use of medical jargon kicks into high gear. Snippets of Madame Psychosis's broadcast, in which she lists medical disorders like enuresis, hyperkeratosis, and hydrocephalus, interrupt the neurologic description of the building. The Student Union takes shape as the narrator tracks a sound engineer's movements through its halls. He "comes in through the south side's acoustic meatus and gets a Millennial Fizzy® out of the vending machine in the sephenoid sinus, then descends creaky back wooden stairs from the Massa Intermedia's Reading Room down to about the Infundibular Recess . . . down past the epiglottal Hillel Club's dark and star-doored HQ, past . . . the airy corpus callosum of 24 high-ceiling tennis courts" (182). This is just one slice of an extensive passage in which human brain anatomy ("sephenoid sinus," "Massa Intermedia," "Infundibular Recess") provides a heuristic for apprehending the urban environment. The narration carefully avoids the language of metaphor or simile here to emphasize that the space is experienced as medical and not just represented as such. The Student Union is not *like* a cerebral cortex; it functionally is one. The body is the vehicle for a conceit that produces a medicalized landscape. Filling out this urban structure are "abundant sulcus-fissures and gyrus-bulges," a balcony "which curves around the midbrain from the inferior frontal sulcus to the parietooccipital sulcus," and a "venous-blue emergency ladder" (186). The descriptions of E.T.A. and M.I.T. establish that, in the world of *Infinite Jest*, one cannot comprehend or inhabit built space unless it is conceived in somatic terms.

Along with the poisoning of earth and soma under O.N.A.N. experialism, the medicalization of the environment suggests the degree to which sickness suffuses contemporary experience. The descriptions attest that, in the twenty-first century, we cannot apprehend clinical bodies and damaged environments in isolation from each other. In one respect, *Infinite Jest*'s corporeal imagination—adduced through

Marathe's wife and the "Québecer" mutants—arises from its environmental imagination of toxification. In another respect, the text's spaces acquire definition through the signifiers of the biomedical body. Environmental critics such as Annette Kolodny and Louise Westling have argued that the gendered body crucially codes the imagination of American landscapes in the twentieth century.[66] Gender, that is, structures an individual's environmental consciousness. *Infinite Jest* introduces a new cultural habit: a biomedical conception of the body performs this structuring function in contemporary fiction. Entangling body and environment conceptually and through narrative causality, the text establishes an environmentality based in the vulnerable, malleable body. It thus depicts the injustices of political, social, and cultural orders of detachment and, as the remainder of the chapter shows, also seeks to remedy them.

AFFECTIVE ITINERARIES

Infinite Jest imagines a contemporary U.S. in which affective and social distancing underpin a spatial politics of detachment that results in environmental and somatic casualties as well as psychic ones. In the text's intricate narration of injury and injustice, Rachel Carson's lessons about the intimacy of human and ecological health intensify, and the vulnerable human body becomes the fulcrum for apprehending our environments and threats to them. The novel overlays earth and soma in one more way: medicalized space indicates the malleability of the city just as toxification signals the porosity of the human body. While built environments within *Infinite Jest* have definite boundaries, when the narrative conceptualizes them in terms of biological systems they become fluid and pliable. The text depicts E.T.A. and M.I.T.'s Student Union through physiological systems of conduction: the heart and lungs that carry blood and air and the brain that carries neural signals, respectively. Conducting flows of air, trash, and energy, these spaces are both rigid and malleable. The novel's mode of describing these and other spaces presses this point. Descriptions are not static in *Infinite Jest*; rather, the narrator constructs environments while moving through them. Building the fluid city through mobile description, *Infinite Jest* introduces the possibility that ways of being in space could remedy detachment. Might avoiding solipsism, detachment, and the anhedonic's "radical abstracting of everything" only be a matter of getting outside and exposing ourselves to the emotional transactions that cities can facilitate (693)?

Edward Soja provides grounds for believing that moving through space can counteract detachment when he argues that spatial manipulation does not always produce distance. He proposes that "to be human is not only to create distances but to attempt to cross them, to transform primal distance through intentionality, emotion, involvement, attachment. . . . In what may be the most basic dialectic in human existence, the primal setting at a distance is meaningless . . . without its negation: the creation of meaning through relations with the world."[67] More than any other character in *Infinite Jest*, Mario Incandenza performs Soja's final dictum: he adds emotional content to life when he moves out into the world. Indeed, the middle Incandenza child models an affective itinerary away from detachment by putting his unsteady feet on the ground; traversing spatial, socioeconomic, and diegetic gulfs; and tapping into others' affect.

Mario continues a robust tradition of urban *flânerie*. As with other literary peripatetics, the uniquely bodied Incandenza covers ground as a way to follow that Forsterian demand, "Only connect."[68] *Flânerie* can fulfill the desire for connection because, in Buell's interpretation, it enlarges "the borders of personal identity such that consciousness extends itself from its perpetually guarded sanctuary into a state of vulnerable, porous transpersonal reciprocity with people and with place."[69] Tracking the roving city dweller through modern literature and cultural theory, Buell argues that this figure exposes himself to the environmental and social vitality of his milieu and thereby acquires an identity based in place. Buell's *flâneur* amends the idea of the early twentieth-century metropolitan that Georg Simmel sketches in "The Metropolis and Mental Life" (1903). Simmel's premise that emotional life in the provinces was as fertile as its soil is perhaps quaint, but his view of the potential risks and payoffs of exposure to teeming urbanity is instructive for analyzing the affective promises of urban space. According to the sociologist, the stable cast of inhabitants of small towns knew one another intimately through quotidian interactions. In the modern city, with its shifting demographics, visual and aural stimuli, and rapid flows of communication and commerce, psychic life necessarily mutates. This mutation amounts to a protective adaptation, what Simmel calls "reserve": "instead of reacting emotionally, the metropolitan type reacts primarily in a rational manner . . . [He] is moved to a sphere of mental activity which is least sensitive and which is furthest removed from the depths of personality," his own and others'.[70] Without "distantiation and deflection," "this type of life could not be carried on at all;" without reserve, misanthropy and even hostility would overwhelm the city dweller.[71] If Simmel recovers reserve, distantiation, and deflection as strategies for surviving city life, Buell directs us to "metropolitan types" who have found this mode of existence stultifying and lived out alternatives. While perhaps faithful to Simmel's world, rationalization has taken a toll on late twentieth-century urbanites. Media and technology have intensified and spread, and detachment and closeted self-absorption are the frightening ends of reserve. In short, the survival strategies that Simmel identifies have become pathological. Urban existence in *Infinite Jest* still produces the "sphere of indifference" that Simmel condemns.[72]

To portray this indifference and imagine alternatives to it, *Infinite Jest* includes both extreme versions of Simmel's metropolitans—that is, people characterized by alienated self-gratification, apathy, and solipsism—and Buell's *flâneurs*—those searching for porous transpersonality. Mario is the latter type, a wanderer who goes out into the street in order to get out of himself. Through this character, the novel contemplates a positive form of vulnerability and exposure that complements the ecologically and somatically damaging dispositions that this chapter has detailed thus far. Somatic vulnerability is undoubtedly harmful, as the scenarios of ecological toxification demonstrate, but it also announces our status as citizens of the more-than-human world and directs attention to systemic political, social, and environmental detachment. Through Mario's movements through space, the text imagines vulnerability and porosity as preconditions for "transpersonal reciprocity" and affective engagement.[73]

Mario's home, E.T.A., exemplifies the "perpetually guarded sanctuary" of individual consciousness that coalesces in metropolitan modernity.[74] Raised from its surroundings, the school promises to insulate young acolytes from life off the tennis court. Monomaniacal focus on tennis justifies a detached disposition that the text otherwise casts as pathological through Avril's debilitating agoraphobia. Athletic director Gerhardt Schtitt informs his pupils that they have the choice to perpetuate their insular state by creating a kingdom of two on the tennis court. In German-inflected prose, Schtitt offers, "'it can be arranged for you gentlemen not to leave, ever here, this world inside the lines of court. You know. Can stay here until there is citizenship. Right here. . . . Or else leave here into large external world where is cold and pain without purpose or tool'" (460). Tennis awakens the young players to networks of interdependence, as Schtitt earlier explains to Mario (81–84), but it is ultimately isolating because the young star "occur[s] as a player" in his own head (461). Total commitment to tennis entails closing oneself off to that "large external world" to which Schtitt alludes, a world that brings anguish and failure but also "access to other selves."[75]

Mario's laundry list of disabilities places him on the periphery of his peers' athletic ambitions as well as enabling him to chart other trajectories for himself that are creatively and affectively rich. As Emily Russell points out, he is one of the "assembled" bodies through which *Infinite Jest* establishes "an ethical model of interdependence."[76] He has a gigantic head, abnormally long fingers, atrophied arms, and block feet that result in "bradypedestrianism" (or slowness) to match his mild cognitive retardation (313). The "lordosis [curvature] in his lower spine" causes him to list forward and "move in the sort of lurchy half-stumble of a vaudeville inebriate" (313). Familial dysautonomia (lack of sensitivity to pain) and the unfortunately matched conditions of poor night vision and insomnia top off the impairments that, in fact, extend his affective reach and make him the most emotionally open character in *Infinite Jest*. Mario's capacity to feel correlates to his status as "a fanatical listener/observer" (189). "One of the advantages of being visibly damaged," the narrative informs,

> is that people can sometimes forget you're there, even when they're interfacing with you. You almost get to eavesdrop. It's almost like they're like: If nobody's really in there, there's nothing to be shy about. That's why bullshit often tends to drop away around damaged listeners,

deep beliefs revealed, diary-type private reveries indulged out loud; and, listening, the beaming and bradykinetic boy gets to forge an interpersonal connection he knows only he can truly feel, here. (80)

Undoubtedly, the connection here is unidirectional. That is, Mario's interlocutors treat the disabled as ciphers, and their callousness is the precondition for this one-sided relationship. Russell is correct that, here and elsewhere, the novel offers up the disabled body as a site of furtive authenticity and thereby plays into an age-old practice of turning damaged bodies into passive pawns in others' games of sympathy.[77] Russell's provocative reading, however, overlooks the fact that this passage is not about how pity accrues to Mario but about the boy's own desires. While his conversations are not the ideal of "transpersonal reciprocity" of which Buell writes, Mario nonetheless yearns to "interface," the novel's argot for "converse," because it enriches him emotionally even if it does not alter his interlocutor's view of him as receptacle. Mario demonstrates his own capacity for understanding others despite their detachment when he puts himself in their shoes: "It's almost like they're like . . ."[78]

Mario's attentive listening is not only an affective adhesive. It also bridges narrative distances, fastening together two of *Infinite Jest*'s plots. The text usually transitions between plotlines with blank spaces between paragraphs or section breaks that announce the year that the subsequent action takes place. Mario's listening first interrupts this pattern and offers stream of audition as a complement to the more familiar stream of consciousness narrative mode. Unable to read, Mario loves video and radio entertainments and is particularly devoted to Joelle van Dyne's "rabidly popular" Madame Psychosis Hour (182). His listening is the link between two narratives that would otherwise be separated by a section break. Snippets of Joelle's radio program, her sound engineer's observations, and a description of M.I.T. give way to the Headmaster's House and Incandenza household without the gap in text (187). This unexpected transition only makes sense once the narrator places Mario "sitting right up close to one of the [radio] speakers with his head cocked dog-like, listening, staring into that special pocket of near-middle distance reserved for the serious listener," and Madame Psychosis's words echo through the

Headmaster's House (189). In this respect, Mario crosses over diegetic as well as emotional chasms.

Along with his habit of careful listening, Mario's observant wandering offers an escape route from the material and affective detachments that compromise sociality and environmentality. He thus instances, in a redemptive register, the novel's case that emotional and spatial relations are interdependent. Despite the physical barriers to his getting out into the world, Mario itches to venture beyond the heart-shaped but heartless E.T.A. On this point, Ennet House is again the academy's mirror image: feeling suffuses the halfway house whereas the school is devoid of it. When he leaves E.T.A., Mario defers to Avril's fears for his safety and limits his range to the Enfield Marine (E.M.) campus on which Ennet House sits. In a bout of insomnia brought on by concern that he's "having a harder and harder time reading Hal's states of mind or whether he's in good spirits," Mario totters down the hill (590). "He likes the E.M. grounds at night because the different brick houses' window-light is yellow lamplight and he can see people on the ground floors all together playing cards or talking or watching TP [i.e., videos]. . . . And a lot of the people in the different brick houses are damaged or askew and lean hard to one side or are twisted into themselves, through the windows, and he can feel his heart going out into the world through them, which is good for insomnia" (590). Simply observing his neighbors exposes Mario to an affective range unavailable at home. The "damaged" but recovering inhabitants of Ennet House are discouraged from censoring reality through irony and apathy as the tennis boys do. "The Headmistress is kind to the people and the people cry in front of each other. The inside of it smells like an ashtray, but Mario's felt good both times in Ennet's [sic] House because it's very real; people are crying and making noise and getting less unhappy" (591). The narrative indicates the sincerity of this moment and brings the reader closer to Mario by signaling that he is the focalizer with the misnomer, "Ennet's House." Mario takes in others' feelings as if breathing in smoke and more clearly apprehends E.T.A.'s emotional sterility. He then uses those expressions to enhance his own emotional repertoire and rest easier. Mario's spatial positioning inverts the anhedonic's: he gets on the ground, gives himself a location, and is thereby *able* to identify. He enlivens his affective being by foraying into built environments in which he encounters others and expands himself. In this respect, he shores up Buell's case that the outcome of urban wandering is the realization that "personhood develops through semifortuitous confluence of people with each other and with physical environment, each shaping the other simultaneously."[79] Mario's journey and Buell's observation help put a finer point on the relationship between body, environment, and affect in *Infinite Jest*: just as body and environment take shape through each other, affective identity only takes shape when one shares others' physical and emotional environs.

Mario's travels provide an at least temporary egress from emotional sterility. But does he model a generalizable antidote to solipsism and ethical detachment? Is his walking a wide-reaching "*therapeutics for deteriorating social relations*," as Michel de Certeau attests in his philosophy of walking?[80] Mario's wandering certainly supplements the meager diet of emotions that he receives at E.T.A. Just as he peers through windows to widen his emotional vistas, he serves as a window onto the relays between space and affect that *Ecosickness in Contemporary U.S. Fiction* elaborates. When read alongside "ecological gerrymandering," toxification, and the novel's medicalization of space, Mario's case supports one of this chapter's claims: that environmental positions—here, one's location in space—provide ontological grounding for affective relations in the twenty-first century. Though Mario brings into relief how anemic these relations are and plainly states a desire for an alternative, his procedure for countering detachment is not the novel's only or most productive one. The conditions that enable Mario's affective encounters make his case exceptional. That is to say, physical disability is the condition of possibility for these affective strolls. While almost all bodies in *Infinite Jest* are exceptional, the extent of Mario's afflictions marks him as a limit case of survivable damage, and few others can walk in his shoes. His somatic specificity undoubtedly makes him the locus of fellow feeling in the novel, but, as Russell contends, it makes him "the ultimate sympathetic object" as well.[81] Russell presents the problem that, even as *Infinite Jest* invites us to feel sympathy for Mario, it suggests that sympathy is tantamount to pity and puts distance between self and other.[82] Sympathy is, in effect, a detached disposition and therefore does not engender radical connection. While Mario's longing to traverse emotional gulfs is sincere and successful on his part, this desire is one that, in my analysis, disgust can more adequately fulfill for others. This is because disgust does not simply place the person feeling the emotion at a distance from the object but also threatens his own ontological status. To understand this mechanism, we must consider how *Infinite Jest*'s depiction of the effects of body-environment interdependence itself generates affect. The patently spatial and yet fluid qualities of disgust are key to how this affect channels the individual outside of the self and toward reciprocal social and ethical involvement.

HOW TO DO THINGS WITH DISGUST

Simmel's account of metropolitan life is negatively charged. "Unthinkable mental condition[s]" arise from "unceasing external contact"; "unwished-for suggestions" from fellow city dwellers are "unbearable."[83] This negativity brings up a concern with celebrating porosity: how much affective input can one absorb before the self starts to burst? Simmel locates this problem in the historical process of modern urbanization. Writing a hundred years later, David Palumbo-Liu argues that the invasion of otherness intensifies as social and other media plug us into global networks pulsing with others' affects.[84] Coping with virtual and physical propinquity to others, we risk slipping into repulsion or even hostility. The detached position that we cultivate to manage such overexposure can carry an "overtone of concealed aversion."[85] Simmel and Palumbo-Liu shine light on an underside of the interpersonal exchanges that Mario seeks in his wanderings: the dissolution of the self might be the outcome of permeability to human and more-than-human otherness. And, as we saw with *The Echo Maker* in chapter 3, this threat to the self can activate compensatory negative affects such as paranoia. In Wallace's novel disgust is the form that affective porosity takes as well as its outcome, and it offers a seemingly destructive response to the realization that we are vulnerable to that which surrounds us. It is a *form* of affective porosity because it is one of the primary transmissible affects; "disgust, being, so to speak, contagious, contaminates its subject as well."[86] It is an *outcome* because awareness of vulnerability incites the aversion response.

As we saw earlier, WWF and Plane Stupid trust that the porosity and mutability of bodies elicit disgust at the same time as this affect incites responsibility for an endangered planet. *Infinite Jest* keys this vulnerability to clinically and anatomically detailed bodies interacting with medicalized environments. In what follows I show how the novel assigns disgust as the affective correlate to a medicalized environmental consciousness. While revealing our susceptibility to our environments and to others, however, disgust does not thereby exacerbate detachment. Rather, it is an unlikely emotional pathway to involvement in the world beyond the self. Putting disgust out there as a means of social and environmental engagement, the novel revalues an emotion that is so easily maligned—if addressed at all—in aesthetic and ethical thought. Concentrating on *Infinite Jest*'s style and the mechanics of disgust, I establish that Wallace ultimately approaches a sick world that is out of joint through disgust. In this way, the novel reconceives this affect as a relation that balances detachment from the world with excessive attachment to the self and accessories to solipsism.

"Balanced" is not an adjective one usually assigns to Wallace's style. Indeed, excess is one of *Infinite Jest*'s motivations and signatures. One of the novel's young characters expresses a wish that we might easily attribute to his creator. As the boys of E.T.A. kvetch about their physical fatigue, they reach the limits of the English language:

"So tired it's out of *tired*'s word-range." Pemulis says. "*Tired* just doesn't do it."

"Exhausted, shot, depleted," says Jim Struck . . .

"None even come close, the words."

"Word-inflation," Stice says . . . "Bigger and better. Good greater greatest totally great. Hyperbolic and hyperbolicker. Like grade-inflation." . . .

"Phrases and clauses and models and structures," Troeltsch says . . .

"We need an inflation-generative grammar." (100)

Infinite Jest's heft and proliferative aesthetic suggest that Wallace enlarged the "phrases and clauses and models and structures" of the novel form to match the content of his story. Reviewing the book for the *New York Times*, Michiko Kakutani castigates Wallace for his "inflation-generative" style. To her mind, Wallace is a "word machine" looking to show off, and *Infinite Jest* is excessively loose and baggy. The novel is "a big psychedelic jumble of characters, anecdotes, jokes, soliloquies, reminiscences and footnotes, uproarious and mind-boggling, but also arbitrary and self-indulgent."[87] Kakutani slams Wallace for his evident lack of control. The laudable novelist, she intimates, must make measured choices. An ounce of control must counterweigh excess.[88] The author must decide when to rein herself in and when to run wild.

Questions of style preoccupied—really, obsessed—Wallace. In D. T. Max's account of Wallace's last years, the novelist's struggle to produce a formally distinct follow-up to *Infinite Jest* intensified the severe depression in which he hanged himself on 12 September 2008. Max quotes a 16 July 2005 email to Jonathan Franzen in which Wallace confesses to his impatience with his signature style: "I am tired of myself, it seems: tired of my thoughts, associations, syntax, various verbal habits that have gone from discovery to technique to tic."[89] Wallace has lost control of a voice that is now entangled and ultimately coextensive with his identity. Reading *Infinite Jest* through the optic of disgust, however, we see what Wallace and Kakutani miss in their assessments. *Infinite Jest*'s disgust idiom, one of the novel's "inflation-generative" features, raises the possibility that excess might be a peculiar form of control.

Ennet House resident Ken Erdedy suggests as much. He trusts that by hitting "Bottom" he can extricate himself from his addiction. To do this, he undertakes a project of excess and induces self-disgust. The narrator describes Erdedy's mission:

> He'd smoke his way through thirty high-grade grams [of marijuana] a day ... an insane and deliberately unpleasant amount. ... But he would force himself to do it anyway. He would smoke it all even if he didn't want it. Even if it started to make him dizzy and ill. He would use discipline, persistence and will and make the whole

> experience so unpleasant, so debased and debauched and unpleasant, that his behavior would be henceforward modified. . . . He'd cure himself by excess. (22)

Erdedy's therapy builds off of the Alcoholics Anonymous principle that the addict must hit the lowest low before she can rehabilitate and surmount her addiction. His program also reveals the expected intimacy of excess and disgust: going too far sets one on a path to repulsion. But here, unlike in cases where we come face to face with putrid meat or a rotting cadaver, the subject and object of the affect merge. Erdedy plans to feel aversion in the hopes that aversion will turn back on him as a sort of homeopathic treatment. The final line here, "He'd cure himself by excess," introduces a question germane to my analysis: can affective and aesthetic excesses—in particular, the disgusting—effectively counteract the afflictions of detachment that are endemic to the contemporary U.S.?

Infinite Jest is replete with characters whose potent, uncontrollable aversions to people, to secrecy, to walls, to bugs, to dirt, and to much else structure their lives. Readers first meet Orin, the oldest Incandenza son, in a scene that elaborates the habits he develops due to his cockroach phobia. He reflects on the first and only time he made the mistake of squashing one: "There's still material from that one time in the [shower] tile-grouting. It seems unremovable. Roach-innards. Sickening" (45). Orin's behaviors and surroundings are shaped by his avoidance and containment of the roach invasions. He takes scalding hot showers, and his home fills with the glass cups that he uses to capture and asphyxiate the critters. To call *Infinite Jest* a novel of addiction is incomplete, then, as it is just as much one of phobic aversion. If the ethical upshot of addiction is distantiation and callousness, what is the ethical upshot of an affect of aversion like disgust?

The text concentrates disgust in descriptions of the somatic effects of environmental reconfiguration and of poisoning through drugs. Only by diving into (or maybe subjecting herself to) the many pages of *Infinite Jest* in which disgust is prominent can the reader fully experience this emotion's effects. Here a few vignettes must stand in for total immersion. I first return to and expand on a passage examined earlier. As Marathe enumerates Gertraude's medical abnormalities, her body violates its bounds to incite the reader's revulsion:

> "She had no skull, this woman.... Without the confinement of the metal hat the head hung from the shoulders like the half-filled balloon or empty bag, the eyes and oral cavity greatly distended from this hanging, and sounds exiting this cavity which were difficult to listen.... Her head it had also neither muscles nor nerves.... There was the trouble of the digestive tracking. There were seizures also. There were progressive decays of circulation and vessel, which calls itself restenosis. There were the more than standard accepted amounts of eyes and cavities in many different stages of development upon different parts of the body. There were the fugue states and rages and frequency of coma.... Worst for choosing to love was the cerebro-and-spinal fluids which dribbled at all times from her distending oral cavity." (778–79)

Marathe's wife is an exemplary figure of excess: her body challenges the limits of survival because it exceeds its own material limits. An excess that the text emphasizes through anaphora ("There were . . . There were . . . There were . . .") and repetition ("cavity," "cavities"). This body is unruly and uncontainable. It is thus abject in Julia Kristeva's simplest definition of this state: that which "disturbs identity, system, order. What does not respect bodies, positions, rules. The in-between, the ambiguous, the composite."[90] The body here does not "stray" onto "the territories of *animal*," as Kristeva writes and as WWF imagines with its climate-altered fishman; it strays beyond the concept of body as that which has determinable boundaries.[91] Gertraude's head hangs over, its parts spill out, "sounds exit" unbidden, and substances flow where they should not.

Leaking fluids carry us to another passage in *Infinite Jest* that centers on the boundless body. In this scene, addict "Poor" Tony Krause's degrading withdrawal from opiates forces him to set up residence in a new Empire Displacement Co. dumpster and then in the bathroom of the Armenian Foundation Library:

> His nose ran like twin spigots and the output had a yellow-green tinge he didn't think looked promising at *all*. There was an uncomely dry-rot

smell about him that even he could smell.... Fluids of varying consistency began to pour w/o advance notice from several openings. Then of course they stayed there, the fluids, on the summer dumpster's iron floor.... Poor Tony Krause sat on the insulated toilet in the domesticated stall all day and night, alternately swilling and gushing.... Time spread him and entered him roughly and had its way and left him again in the form of endless gushing liquid shit that he could not flush enough to keep up with. (301–3)

The horrors of bodily chemistry play out here: no longer injecting the Substance, Poor Tony's body concocts and ejects the substances that both originate in the self and are grossly foreign to it. This uncontrollable self-defilement is only more disgusting for the obsessively hygienic Tony. His withdrawal lends truth to Kristeva's point that the abject comes from within as much as it enters from without. "It is as if the skin, a fragile container, no longer guaranteed the integrity of one's 'own and clean self' but . . . gave way before the dejection of its contents."[92] Poor Tony does not find this mortifying experience reassuring, and, as rapacious time dilates, the reader too wants to get out of the confined space of the bathroom stall in which his personal torment unfolds.

In these and similar passages, characters' bodies serve as vehicles for narrative disgust. In particular, Gertraude and Tony Krause engulf the reader because they exemplify "violations of the exterior envelope of the body," one of the nine primary elicitors of disgust that the Plane Stupid "Polar Bears" campaign also deploys.[93] Bodies bursting through their own boundaries materialize the many ways in which we are living in "chemically troubled times" (151). The literary works of these times, and ecosickness narratives particularly, rarely represent the body in ways that don't invite some measure of disgust.[94] Disgust is a powerful strategy for making bodies physical even as it foregrounds the indeterminacy of the body's limits and thus threatens the self. But this threat is just what we need, Wallace's text shows. On my account, "the proximity of the bodies of others" is not only "read as the cause of 'our sickness,'" as Charles Darwin would have it;[95] this proximity is read as the potential remedy for the sicknesses of social and environmental detachment.

The novel focuses disgust on the contaminated bodies that processes of detachment have produced as a way to attune us to the material impossibility of total detachment.

In *Infinite Jest* readers get uncomfortably close to Gertraude's and Tony Krause's unpredictable bodies. They present cases where poisoning, whether from environmental toxins or drug abuse, renders the body radically unfamiliar. A lexicon of disproportion—"more than accepted amounts," "rages and frequency," "could not . . . keep up with"—heralds this strangeness. By making the body strange, Wallace reaches toward one of the goals that he assigns to contemporary fiction. In "E Unibus Pluram" he observes that "today's most ambitious Realist fiction is going about trying to *make the familiar strange*" at a historical moment when "we can eat Tex-Mex with chopsticks while listening to reggae and watching a Soviet-satellite newscast of the Berlin Wall's fall—i.e., when damn near *everything* presents itself as familiar" (52). Like *The Echo Maker*'s Gerald Weber, who laments that "progress would at last render every place terminally familiar,"[96] Wallace finds that the globalization of taste and of communication networks has the same effect on experience writ large. The novelist must reverse the trend toward excessive familiarization and make the ordinary alarmingly strange. Several scholars remark upon *Infinite Jest*'s literary efforts at defamiliarization, largely focusing on recognizably postmodernist strategies that violate realist expectations.[97] While I agree that multiperspectival, radically nonlinear, and deliberately inconsistent narration destabilizes the reader accustomed to domestic and social realism, *Infinite Jest* does not solely aim to swerve literary history and test readers' patience and interpretive skills. As I've already noted, Wallace strives to "'author things that both restructure worlds and make living people feel stuff.'"[98] To this end, he takes up the manifold project of recoding space, reorienting relations between self and human and nonhuman others, and experimenting with an affect that "radically disturbs different relations of proximity."[99]

Disgust, a peculiarly embodied emotion, puts up a dam in the onrush of the overly familiar. It directs attention to what was previously ignorable—in this novel the somatic, ecological, and social injuries of detachment—and manages response to disquieting realities. Rather than distance a person from the outside world, as we might expect, the defamiliarizing affect of disgust reattaches her to it.[100] Germane ideas about the workings of disgust from philosophy, aesthetic and cultural theory, and psychology ground my investigation into whether a negative emotion like disgust can promote the kind of affective and ethical involvement toward which *Infinite Jest* aims.

Erdedy's push to cure his addiction through extreme unpleasantness is premised on the belief that disgust is compelling because it is such a visceral emotion. Directing enough disgust at himself, his thinking goes, he cannot help but eradicate the source of disgust. Philosopher William Ian Miller's claims for disgust lend support to Erdedy's plan.[101] Miller contends that the emotion

> has certain virtues for voicing moral assertions. It signals seriousness, commitment, indisputability, presentness and reality. . . . Our moral discourse suggests we are surer of our judgments when recognizing the bad and the ugly than the good and the beautiful. And that's at least partly because disgust (which is the means by which we commonly feel the bad and the ugly) has the look of veracity about it. It is low and without pretense. We thus feel it trustworthy, even though we know it draws things into its domain that should give us pause. The disgust idiom puts our body behind our words, pledges it as security to make our words something more than *mere* words.[102]

Unambiguous and undeniable, disgust has a reliability that feelings of unequivocal desire can lack. Though Miller contrasts disgust to the beautiful, disgust, like Kant's beautiful, demands assent, but for different reasons.[103] Our reaction is so much in the body that we believe, perhaps naively, that prejudice or social norms have not contaminated it. The trustworthiness of this emotion makes it especially powerful in a novel where questions of fact—is Joelle perfect or deformed? Is "Infinite Jest" buried in Jim Incandenza's skull? Does Don Gately survive?—are constantly running through the reader's brain. Yet disgust also "seeks to include or draw others *into* its exclusion of its object, enabling a strange kind of sociability."[104] In other words, disgust is an affect of inclusive exclusion; it generates community around its aversion to objects and others. The emotion can therefore be a powerful corrective to the solipsizing allures of contemporary life, everything from *M*A*S*H** to meth, that *Infinite Jest* inspects. Media-induced stupefaction has become a solitary pleasure and has skewed judgment in Wallace's imagined twenty-first-century U.S. Disgust might reorient attention through its embodied trustworthiness and even draw individuals into a community.

For these reasons, Sianne Ngai puts disgust at a threshold point in her account of "ugly feelings." As the next chapter on *Almanac of the Dead* demonstrates, some affects—anxiety, for example—are anticipatory and suspend agency. By contrast, disgust's immediacy, intensity, and certainty put us on the cusp of "more instrumental or politically efficacious emotions."[105] Disgust is politically powerful by Ngai's reckoning because it is unignorable. We must ask, though: is this always the case? What if we cannot stand to examine the offending source? We may be certain about how we feel and still not want to dwell in that certainty. Architectural critic Mark Cousins's treatise on the ugly introduces this possibility into the conversation. The disgusting object comes at us as a threat; on this philosophers from Kant to Jacques Derrida agree. In the face of this threat, we have two choices according to Cousins: "to destroy the object, or to abandon the position of the subject. Since the former is rarely within our power, the latter becomes a habit. The confrontation with the ugly object involves a whole scheme of *turning away*."[106] Accordingly, even though the disgust response may be unequivocal, the object also produces a desire to avert our eyes, to put space between ourselves and the imposing elicitor of disgust. In this respect, then, disgust is a particularly spatial emotion, as Kristeva also argues.[107] The history of thought on disgust bears out this observation. Beginning at least with Thomas Hobbes, the philosophy of emotion has categorized it as a "retiring" affect that moves us "fromward."[108] Given this, disgust, while initially a powerful, unignorable draw on our attention, mutates into an instrument of disattention that promotes the socially and environmentally damaging detachment that animates *Infinite Jest*'s storyworld.

Our own experience of disgust might substantiate this idea that we want to turn away from (and in) disgust. But we have also surely experienced the opposite response: even after looking away, we turn back for another look at the source of aversion, peeking between the fingers that shield our eyes. Perhaps even involuntarily, as Miller notes.[109] We waver between repulsion and attraction. This vacillation is central to Pierre Bourdieu's definition of disgust. Following Derrida, he defines it as "the ambivalent experience of the horrible seduction of the disgusting and enjoyment" and as capable of "driving one irresistibly towards consumption."[110] Or, to cite another affect theorist who updates this lineage, it "brings the body perilously close to an object only then to pull away from the object in the registering of the proximity as an offence."[111]

Conflicting drives to attend to and turn away from the repulsive constitute the emotion of disgust. Put in dialogue with these other thinkers of disgust, then, Miller's point that the emotion is "indisputable" calls for nuance. The *content* of disgust may be undeniable and therefore morally powerful in that its source is determinable and the affect speaks through the body. Yet the *form* of disgust is vacillatory. Spatial oscillation, toward and away, goes along with an ontological oscillation between what David Pole terms "dissociation" and bonding.[112] At first, a "posture of dissociation . . . announces [disgusting objects'] unconnectedness with me," but then this easy interpretation becomes less tenable.[113] This is because "that announcement . . . is always false, for otherwise they would never disgust me. Disgusting things, in order to be physically disgusting, must either be able to be seen as simultaneously appetizing, desirable, or worse still, already be part of me."[114] In a psychoanalytic register, Kristeva also makes the point that what appears to be an affect that unequivocally distances and distinguishes self from other instead shows the constitution of the self by that which surrounds it.[115] It shows the body to be enmeshed in a mangle of "sticky" relations with its world.[116]

This is one respect in which disgust is a crucial environmental affect in an age of pervasive detachment: that which constitutes one's world extends beyond other humans to include inanimate objects and, most importantly, the more-than-human realm. Moreover, the conflicting urges to attend to and to turn away from—to identify with and to distinguish oneself from—the repulsive are vital to an environmental discourse invested in attention and awareness, as my analysis of wonder in the previous chapter suggests. In chapter 3 I surveyed the many pronouncements of a linear relationship between attention and commitment to planetary care within environmental thought. Environmental thinkers, including Edward Abbey, Rachel Carson, Mitchell Thomashow, and Scott Slovic, stake their projects in the bedrock of ecological attention. The alternative of disattending is anathema to environmental ethics, as it arguably is to ethics more generally. Part of *Infinite Jest*'s project is to prod readers to do an about-face in their habit of turning away from their own and others' realities. Erdedy's trip to the Bottom can set off a parallel trip out of detachment if the excess indulgence cures his addiction, but it's first a trip away from the world: sequestered in his bedroom, Erdedy will be "unreachable for days" and request that the pot courier throw him the goods while he "would from a distance toss back to her the $1250 U.S." (20, 23). In detailing addiction and the detachments of experialism and ecological gerrymandering, Wallace certainly writes from an antidetachment platform. The issue here is that *Infinite Jest*'s prevalent use of the disgusting might work at cross-purposes with its condemnation of disattention and of detachment. How can a text that deploys disgust ensure that readers don't just disengage?

One way Wallace skirts the danger of inattention is by formally coding disgust's tendency to vacillate in the interplay between hyperbole and mimesis, by which I mean commitment to the order of relations outside the text. Along with its nonlinear narrative, tortuous syntax, and endnotes, the novel revises our reading strategies through this dynamic. In the selections from *Infinite Jest* that I have analyzed thus far, the disgusting arises from observable biological phenomena. Specialized medical jargon punctuates the text when it portrays somatic vulnerability. Committed to a degree of mimesis, these scenes demand readers' attention. They call up our knowledge that something akin to what we are reading could and indeed does occur to people out there and just might occur to us. Marathe's medical blazon gives life to the horrific effects of poisoning from environmental toxins, and Krause's thoughts conduct readers through the phases of heroin withdrawal. Neither of these scenes would cohere without either a measure of referentiality or a great deal of narrative scaffolding. That said, *Infinite Jest* is by no means strictly faithful to evidence from lived experience of these disorders. The narrative hyperbolizes scenarios of environmental and physiological mutation and injury and thereby defamiliarizes the body, the environments in which it withers or thrives, and the interactions between them.

Because it is grounded in biological specificity, the interplay between mimesis and hyperbole energizes what Paul Ricoeur calls "the shock of the possible" in fiction.[117] To take Gertraude's case, *Infinite Jest* derives her ailments from the medical reference books one might well find at the M.I.T. library. Yet the narrative diverges from medical paradigms as these injuries pile up. Really, the reader must ask, could she survive them all? As Krause's trip "way down the deserted corridor of Withdrawal" spreads out over six pages of tight print and the narrator itemizes every fluid that could flow from a body's hidden orifices, the text amplifies addiction's dangers (300). Likewise, the Concavity's rapid fluctuations between barrenness and fertility violate mimetic expectations. Through disgust, then, *Infinite Jest* toys with degrees of closeness to and distance from the threats to somatic and environmental integrity that produce the emotion. Wallace's novel draws out one of the challenges of narrating ecosickness as well as its importance. As *Ecosickness in Contemporary U.S. Fiction* has argued, in order to depict somatic and environmental injury as shared phenomena, recent fiction shows the interdependence of human biological, ecological, and social systems without resorting to linear causal models. *Infinite Jest* exposes another feature of this literature: authors must weigh how faithful they'll be to medical and environmental realities as they refract conditions of alienation and toxification and activate responses to them. *The Echo Maker* approaches this challenge by toggling between the familiar and the strange to elicit wonder and show the affect's limits to care. In a parallel, though not equivalent, way, *Infinite Jest* balances referentiality and hyperbole, capturing readers' attention, poking their affective core, and daring them to look away.

Just as the novel must balance mimesis and hyperbolic invention, it also must balance overstimulation and being "divorced from all stimulus" (142).[118] At either of these two poles, we risk becoming numb to the world outside the self. Overstimulated, we attend fixedly, even obsessively, to an often damaging substance, habit, or sense of self. Within the text this habit manifests in O.N.A.N. society's absorption in drugs and entertainment as well as in characters' solipsism. Understimulated, we are apathetic, affectless, and cannot invest in anything beyond ourselves. (Of course these poles are traversable: *Infinite Jest* presents several cases where overstimulation shades into understimulation.) Wallace addressed this need for balance and urged openness to attention in his address to the 2005 graduating class at Kenyon College and in the posthumously published *The Pale King* (2011). In his speech he advised his young listeners to reconceive freedom as "attention and awareness and discipline, and being able truly to care about other people."[119] *Infinite Jest*'s idiom of disgust activates involvement as a countermeasure to the material and subjective detachments that manifest in contemporary American psychic, political, and social life. A chorus of thinkers about postmodernity has made cognate claims about the solipsism and apathy of late twentieth-century existence. Wallace contributes to this conversation by elaborating a medicalized environmental consciousness in which disgust draws together social, environmental, and aesthetic interventions into these conditions. Through a medicalized form of disgust, *Infinite Jest* modulates the actual and the invented in order to balance overattention and underattention. Disgust, with its dual aspect of drawing us in and pushing us away, satisfies the demand for an even attention that can negotiate self-awareness and awareness of human and more-than-human others. It offers a surprisingly curative invasion, a way to forge bonds.

Negotiating attention to self and to others depends on the kind of reflection that novels invite. The novelistic, in particular, suits what Michael Bell and Jane Thrailkill define as "double-feel": "(1) the enlistment of readers not just to feel emotion for . . . an unreal object [that is, fictional characters and events] . . . but also (2) to engage in this process of reflection or double-feel, to realize one's creative participation in experiencing the text and indeed the world."[120] This process requires the slow unfolding and recursivity of narrative. To varying degrees, all novels afford readers the time and space to look close, pull back, and then return for another glimpse.[121] A markedly recursive novel such as *Infinite Jest* further encourages this behavior through its nonlinear diegesis, self-referentiality, multivocal narration, and endnotes, which demand that one read forward and backward at once. An impetus to reflective reading, the disgusting has the potential to move us from observation to involvement and to enhance "the reader's education *about* and *through* his emotional engagement."[122] This emotion drives us to consider not only the objects, relations, and boundary violations that generate disgust but also how affective bearing alters ethical bearing. If the contemporary ethos in Wallace's novel is to "send from yourself what you hope will not return" (1031*n*168), temporarily but wholeheartedly tramping through the disgusting offers an alternative in which you release such domination and allow the outside to come streaming in. *Infinite Jest* thus positions disgust against self-absorption and environmental, psychological, and social detachment in order to manage interaction between self and world through word.

If AIDS memoirists revalued discord and *The Echo Maker* established paranoia as the shadow of wonder, *Infinite Jest* gives disgust a page in the catalog of environmental affects. It might surprise readers that Wallace's novel enlarges the emotional ambit of environmental thought. After all, unlike the WWF and Plane Stupid campaigns with which this chapter began, and unlike Marge Piercy and Leslie Marmon Silko, the focus of the next chapter, Wallace does not have stature as an environmental thinker, much less activist. Reading *Infinite Jest* through the optic of sickness resurrects questions that have preoccupied ecocriticism: must environmental literature worthy of the name serve environmental*ist* ends? Must it point fingers and provide solutions? Is it enough for literature to raise questions or expose the complexity of interconnection in an eco- and biotechnological age? Or must we walk away with at least a spare blueprint for behavioral and policy changes?

I return to these questions in the conclusion to this book. With respect to *Infinite Jest*, I argue that, in not sketching a precise roadmap for environmentalism, the novel challenges environmental thinkers to excavate the habits of thought and dispositions on which environmental crises rest. It lends support to the idea that novels are still potent tools for thinking through a key question that has animated the genre for centuries: what does it mean to be an embodied, feeling person negotiating both interpersonal relationships and macrosocial forces such as geopolitics and environmental change? As the novel makes damaging social and environmental policies visible through damaged human bodies and depicts affective itineraries toward reattachment, it suggests that proposals for individual and structural reform will fail if we do not address the ethos that underpins the problems those proposals are designed to solve. For Wallace, that troubling ethos is detachment. Developing new energy solutions and combating versions of experialism to protect environmental and bodily health are fool's errands so long as we do not swerve away from the detachment that fuels contemporary U.S. consumption and exploitation.

5

The Anxiety of Intervention in Leslie Marmon Silko and Marge Piercy

Standing in the twenty-first century, it's clear that the "age of anxiety" that W. H. Auden declared in 1947 is still with us. But some things have changed. In the wake of World War II, writers looked to the inventions of total war, notably to nuclear weaponry, when composing their visions of technological anxiety. In the last decades of the millennium, nuclear worry does not disappear, but it cedes some ground to concerns about techniques for radically altering so-called nature. Biotechnologies that transform life take center stage; they change the world not through the spectacular blast but through gradual reconfiguration of life's building blocks.[1] We need only look to Leslie Marmon Silko's career to chart this trajectory from World War II to the new millennium. Her debut novel *Ceremony* (1977) captures the earlier moment: a Laguna Pueblo woman's memory of the test detonations at the Trinity Site in New Mexico establishes prediegetic time, and the plot ends at a uranium mine where the U.S. military sourced raw materials for the bomb. Silko's next novel, *Almanac of the Dead*, was composed throughout the 1980s and published in 1991 and is suffused with biotechnological anxiety. It bears the imprint of its moment in time, depicting biological interventions and management under neoliberalism. Here technoscientific developments such as gene therapy and artificial ecosystems alter the matter of humans and nonhumans. Though the historical positions of *Ceremony* and

Almanac differ, Silko organizes both texts around forms of ecosickness that correlate to the exploitation and dispossession of the poor and Native people.

As her focus on technological anxiety increases with *Almanac*, Silko joins a tradition of women writers imagining apocalypse through biomedical and environmental change. A representative lineage of these fictions stretches from Marge Piercy's *Woman on the Edge of Time* (1976) to Octavia Butler's *Parable of the Sower* (1992) and into the twenty-first century with Jeanette Winterson's *The Stone Gods* (2007) and Margaret Atwood's Madd Addam trilogy (*Oryx and Crake* [2003], *The Year of the Flood* [2009], and *Maddaddam* [2013]). Women's literature of the 1970s and 1980s was depicting the environmental repercussions of biotechnology and the inequities that biotech exacerbates before these problems became a focus of ecocriticism in the 1990s and early 2000s. This chapter focuses on Piercy's and Silko's speculative novels, whose slightly exaggerated presents and invented futures revolve around pervasive conditions of sickness and the structural inequalities of interventions into body and land. *Woman on the Edge of Time* and *Almanac* pointed the way for an environmental studies more attuned to somatic and ecological interventions as shared catastrophes. Read together, they invite theoretical questions about whether deploying the narrative affect of anxiety can help usher in a more just future. Piercy's and Silko's apocalyptic fictions feature anxiety on the menu of affects attached to degradation and sickness. In their books the affect of anxiety is concomitant to rapid, irreversible changes that favor wealthy elites and that emerge from biotechnological research. The question for both texts is the following: Do the anxious apocalypses that they prophesy invite or foreclose their visions for environmental and somatic renewal?

Woman on the Edge of Time was composed at the launch of the biotech revolution and at a moment when modern environmentalism and the spatial marginalization of the urban poor were provoking debates about how land use expresses social and species injustice.[2] Piercy's novel shows not only the ghettoization of the ethnic poor in New York City but also the ghettoization of the female mind and body through psychiatric institutionalization and biomedical intervention. The Chicana protagonist, Consuelo (Connie) Ramos, is medically sedated and imprisoned in an asylum with a diagnosis of schizophrenia. Connie herself can't decide whether to dismiss or accept this label given that she is visited by Luciente, a time traveler from one version of the year 2137. Luciente shows Connie the near-utopian community of Mattapoisett, an agrarian, equalitarian society on Cape Cod that "must fight to exist, to remain in existence, to be the future that happens."[3] This future world is only possible if Connie's generation rises up against the wealthy and powerful to make techno-science accessible to all. The novel therefore does not propound Luddism in response to technologization; in fact, it endorses genetic and mechanical technologies that relieve the burdens of alienated labor and reproduction to make time for self-exploration and pleasure. In *Woman on the Edge of Time* anxiety is not the response to pervasive sickness so much as a form of sickness, and a narrative of restoration is available not despite but because of technological inventions that promote equality.

Almanac reworks Piercy's novel in that it generalizes the imbalances of technoscience, narrating their effects not just on poor women and homosexuals but on the destitute, on Native peoples in the Americas, and on all people of color. However, its organizing principles of apocalyptic anxiety and revolutionary regeneration are in conflict with each other. After a short reading of *Woman on the Edge of Time*, this chapter analyzes Silko's novel set on the cusp of revolution and the restoration of the sick bodies and lands that have been expropriated from indigenous people and the poor. At one level *Almanac* presents body and land as conduits to healing through a rhetoric of merging familiar to environmental thought. However, land and body are also the points of application for unbridled technoscience, and scenarios of bio- and ecotechnological violence produce the novel's anxious tone and interrupt the therapeutic fusing of human and more-than-human. *Woman on the Edge of Time* resolves the conflict between technoscientific progress and social justice by concentrating the narrative in a single protagonist who shuttles between a degraded present and a fully realized—albeit fragile—utopia of technologized agrarianism. In *Almanac*, however, the conflict between technoscience and justice persists, and anxiety is generalized and arises from two narrative strategies: medicalizing tropes that affiliate sick earth and sick body and apocalyptic discourse. This chapter establishes that *Almanac*'s anxious tone erodes the novel's vision of revolutionary healing due to the paradoxes of anxiety. The very emotion that triggers revolutionary fervor about the dangers of emergent technologies in fact suspends the capacity to resist them.

Ultimately, *Almanac* shows that our theories of narrative affect change based on whether they have a subject at their center or are dispersed across a text. Just as *Ecosickness in Contemporary U.S. Fiction* elaborates the environmental affects that contemporary narratives deploy to refract sickness today, this chapter models an ecocriticism that accounts for the full affective spectrum of environmental fiction, from restorative place sense to techno-anxiety.

DISRUPTING THE "PATTERN OF DISEASE"

It is difficult to imagine an apocalyptic narrative that does not aim to generate anxiety. Even in its comic strains, apocalypse puts the reader on edge along with the characters living through the havoc of a ruined world. Piercy's *Woman on the Edge of Time* routes its anxiety through the precarious economic and social conditions of women of color and gay people who have been physically and psychologically ghettoized. Pervasive violence and death; addiction and prostitution; and alienation from place, memory, and family create the novel's apocalyptic texture. Connie is living in what the Mattapoisettans of 2137 dub "the Age of Greed and Waste" (47). According to the novel, citizens of the late twentieth century have not learned the lesson of the Manhattan project: that "technology itself [is] a threat" (48). As environmental cataclysms increase, Americans devise solutions that only produce more ecological upheaval. Luciente, the "sender" from Mattapoisett who reaches out to Connie, relates, "It rained for forty days on the Gulf Coast till most of it floated out to sea . . . The jet stream was forced south. They close to brought on an ice age. There was five years' drought in Australia. Plagues of insects . . . Open your eyes" (89, ellipsis in original). While the technological interventions into climate systems are disastrous, *Woman on the Edge of Time* shows biomedical interventions to be the most apocalyptic, since they maintain the power of white male technoscientific experts and those who consort with them. The destinations for these interventions are the bodies and the minds of people of color and the poor who lash out against their impoverished conditions.

Connie faces an onslaught of state-based surveillance strategies and ways of managing bodies that mimic colonialist discourse and practices. These include imprisonment, social services monitoring, and institutionalization. Incarceration takes many forms in the novel and becomes a metonym for social disintegration and disenfranchisement. Medical imprisonment literalizes this powerlessness. Most of the novel's action takes place during Connie's second forced institutionalization. At Rockover State she becomes the subject of invasive mind control experiments that turn the prisoner-patient into "a walking monster with a little computer inside and a year's supply of dope to keep her stupid" (277). In targeting the "centers of aggression, the primitive emotions run amok," the hospital's neurosurgeons and psychiatrists pacify those patients whose intransigence challenges elite privilege (276). The novel's climax occurs when the doctors manipulate Connie's brain, but this act is only the last in a series of self-imposed and forced medical procedures that *Woman on the Edge of Time* details. Connie's medical history includes a hysterectomy to which she didn't consent, and, throughout her incarceration, she trembles at the threat of electroshock therapy and the neuro-implantation surgery to which the doctors eventually subject her. The narrative presents these experiments in the language of penetration and invasion. "They were running the savage bolts through [the] soul" of Sybil, Connie's confidant in the asylum (103). Piercy alludes to Mary Shelley's *Frankenstein*, but updates it for a biomedical age when needles take the place of bolts and can alter the mind and spirit (186).

Connie resists the male doctor-scientists' mental mutilation by violently appropriating the "powers of life and death" that the experts normally wield (370). If they want to neutralize her through pharmaceutical and medical procedures, she will turn the tables on them by weaponizing pesticides, a chemical cousin of their biomedical tools. In her culminating act of rebellion, "she grabbed up a small bottle [of parathion] and filled it with the brown oil, her hand trembling. Slowly she poured it, holding her breath. Perhaps even coming this close might kill her, but then they were going to kill her anyhow. But this was a weapon, a powerful weapon that came from the same place as the electrodes and the Thorazine and the dialytrode" (357). The military diction here suggests that the promised "thirty-year war" between "those who controlled" and those who are controlled has already begun (190, 357). The war is an outcome of a technological struggle that a Mattapoisettan relates: "at certain cruxes of history . . . forces are in conflict. Technology is imbalanced. Too few have too much power. Alternate futures are equally or almost equally probable" (189, ellipsis in original). In using the scientists' tools against them, Connie shows one way to usher in the uprising that will right the balance of technological power. However, even as *Woman on the Edge of Time* depicts Connie's fierce agency, it does not make her the trailblazer for revolution. Not only do the doctors escape death, the novel ends with a parallax perspective on the narrative through the protagonist's medical chart. According to this document's pathologizing logic, Connie's journeys to Mattapoisett are signs of schizophrenia rather than boosts to her self-esteem and political agency.[4] Even if she is not the insurgent and defender of the future that the Mattapoisettans need, however, Connie channels the utopian impulse. Her initial skepticism toward the socialist society may placate a cynical reader, but she ultimately comes to embrace Luciente's world.

The narrative thus positions Connie on several edges: of time, of neurological manipulation, of environmental cataclysm, and of a revolution that ushers in the preferred future. But the narrative itself is also on the edge: of time—as present and future are suspended within it—and of anxious affect. That is to say that, while *Woman on the Edge of Time* unleashes anxiety with its scenarios of biotechnological oppression, it also retracts it. The "dreadful anxiety" that besets Connie after the implantation surgery is merely a ramped up version of the "general mood" of anxiety within which she lives (296, 373). Piercy's novel does not narrate the conflagration that has reduced the U.S. to ashes by 2137, but the portents of catastrophe and the personal and collective anxiety that imminent mass destruction produces are nonetheless everywhere in Connie's world. Ecological and medical technologists are running wild. The day is described as "bleeding at the edges" (26). The food supply is poisoned with "nitrites, hormone residues, [and] DDT" (46). Vulnerable people are dehumanized, transformed into machines at one turn and animals at another (195). Utopia may elicit anxiety as Fredric Jameson argues,[5] but it's also the case that the conditions that herald the *need* for utopian thought spark an even stronger form of the affect.

Institutionalized gender and sexual oppression, racism, and other forms of material and spiritual impoverishment form the tapestry of anxiety that the novel weaves. Biotechnology is the phenomenon that threads through this cloth. It produces the tone of anxiety in *Woman on the Edge of Time* and, as we will see, in *Almanac* as well. However, in the former book, biotech also releases that anxiety. The novel concurs with Perry Anderson when he declares that utopianism after the dawn of recombinant DNA must contend with this new tool. He urges that "the biological ground can no longer—if it ever could—be left to those devoted or resigned to the established order of capital; it will have to be invested in new ways too."[6] In Piercy's novel the middle managers of the "established order" are medical professionals. The concentration of technological expertise in this one elite group ensures the continued abuse of poor and racialized communities. However, it is this misuse of technoscientific prowess, and not the technoscience *itself*, that Piercy's text indicts. Utopia requires a takeover of biotechnological agendas, not the abolition of them. While anxious affect surrounds biotech interventions, Mattapoisett society shows that inequity does not inhere in the manipulation of life itself.

Without genetic, mechanical, and communication technologies, the Mattapoisett utopia of universal labor would not be possible. Luciente instructs that this liberation depends on all members of the community having access to these tools and not only the powerful. As if teaching a child, she explains to Connie the basic philosophy of technoscientific syncretism: it's a matter of "put[ting] the old good with the new good into a great good" (63). Just as the declining present and the restored future coexist in the diegetic time of Piercy's novel, less sophisticated and advanced technologies coexist in the better world of Mattapoisett. They genetically modify mosquitos to rid them of their poison (89). They familiarize children with "spectroscopes, molecular scanners, gene readers," and other devices to expose them to the work of genetic engineering (125). These innovations ease the burdens of reproduction and agricultural production and free time for the enriching leisure that produces societal stability.[7] With technoscientific inquiry and practice in the hands of any Mattapoisettan who wants them rather than "'huge corporations and the Pentagon,'" public debates precede the development of new genetic interventions (272). They adhere to the precautionary principle, acknowledging that there are limits to their understanding of how natural systems work, and weigh collective benefit against possible risk. Holism guides their efforts to eradicate environmental and human sickness. If technoscientific experts are making Connie's life "into a pattern of disease" through surveillance, surgery, and pharmaceuticals, the Mattapoisettans are monitoring patterns of disease through genetic technologies in order to liberate their society from sickness (19). Ultimately, in *Woman on the Edge of Time*, humans enter mutually salutary relations with the more-than-human world through biotechnological innovation. As we will see, in *Almanac* biotech is inherently iniquitous and interrupts this relation.

Portraying technoscience as a tool both of patriarchal privilege and of liberation from the uneven distribution of opportunity, Piercy's novel defuses the anxieties that it produces. It also mitigates the anxious tone that techno-horror introduces with its narrational strategies. The third-person narration focalized through Connie keeps the reader close to the protagonist throughout her imprisonment and her passing, perhaps hallucinatory, sojourns to Mattapoisett. In one respect this closeness makes the text claustrophobic, akin to the spaces in which Connie is confined. There appears to be no respite from Connie's self-interrogations and hopelessness. "What did she live for?" she wonders in pain. "Protecting Dolly? Could she protect Dolly, really? A fantasy of someday recovering her daughter?" (41). However, with a single protagonist as the narrative center, *Woman on the Edge of Time* reins in the anxiety that I argue is expansive and unremitting in *Almanac*. The focused characterization in Piercy's novel also expresses an enduring trust that human agents determine the future. "'There's always a thing you can deny an oppressor, if only your allegiance,'" Bee instructs Connie. "'Often even with vastly unequal power, you can find or force an opening to fight back'" (322). Though Connie expresses doubt, the reader knows that she holds to this principle. She regards a fellow patient following his neuro-implantation surgery and assures herself, "he was no robot, whatever they thought they had done. She could feel the will burning in him, a will to burst free" (264).

It's on this point that we must turn to *Almanac of the Dead*. For, in Silko's novel, the matter of how human agency endures in the face of technological interventions into body and earth becomes even more urgent. As *Almanac* reworks Piercy's speculative fiction about the inequities of late twentieth-century technoscience, it turns up the dial on anxious apocalypticism and introduces the question whether anxiety can catalyze sociopolitical and environmental restoration.

"A SINGLE CONFIGURATION" OF LAND AND BODY

Showing the late twentieth century to be a terrifying time of interventions into life itself, *Woman on the Edge of Time* and *Almanac of the Dead* both stage conflicts between technoscientific progress and social justice and pin hopes for justice on revolutionary upheaval. The latter work depicts the same matrix of concerns that preoccupies the former: structural racism, pervasive bodily and environmental sickness, the manipulation of space, and biotechnological growth. In Silko's novel these conditions are magnified and the dispossessed are shouting "Enough!" While Piercy's novel has a U.S. focus, *Almanac* traverses the Western hemisphere, and its historical imaginary includes the *longue durée* of slavery and colonialism in the Americas since Columbus's arrival. From the Yukon to the Yucatán and points south, the Americas are a polarized world of privilege and disenfranchisement. Violent exploitation of the poor and of the environment find common cause in the novel's two historical and economic determinants: colonialism and neoliberal capitalism.[8] Privatization, the commodification of body parts and the earth, and the expropriation of land ensure that capital and power accrue to those who already have the most.

Like *Infinite Jest*, *Almanac* is a long speculative fiction whose very heft suggests the enormity of its project to depict pervasive sickness and territorial reconfiguration and to write a way out of them. Silko's novel bears the imprint of its moment in time—from population growth in the southwest U.S. and New Age spirituality to radical environmentalism and the construction of Biosphere 2. Yet it is also undated; its contextualizing allusions point only to an "after"—after the dropping of the atomic bomb, after African independence movements, after Vietnam, after the birth of the internet. In Silko's 1977 novel *Ceremony*, post-traumatic stress disorder and alcoholism are symptomatic of the young Laguna generation's disconnection from place and tribal tradition. In her 1991 book, sickness encompasses drug addiction, disability, and cancer, and afflicts Americans of all ethnicities, races, classes, and genders. In both novels drought and theft of Native lands in the Southwest are metonyms for all anthropogenic threats to the environment. However, between the publication of *Ceremony* and *Almanac*, the narrative template of healing is less assured as anxieties mount about technological interventions that alter the matter of life itself.

The desert territory spanning the U.S.-Mexico border is the geographical center of *Almanac*'s main story about indigenous groups' fight to "bring all the tribal people of the Americas together to retake the land" that Euro-American "Gunadeeyahs," or "Destroyers," stole over centuries of colonization.[9] The novel's more than seventy characters divide into those promoting socioeconomic and environmental injustice and those resisting it. As the book opens, Lecha Cazador, a Yaqui Indian with psychic powers, returns to her twin sister Zeta's ranch on the outskirts of Tucson. Lecha's cancer has brought an end to her nomadism, and she reunites with her family as they run drugs and stockpile arms for the coming revolution. A white recovering addict and sex worker, Seese, assists Lecha in her project to transcribe the Yaqui almanacs, and befriends Sterling, the Laguna gardener on Zeta's property. Through Zeta's involvement in the weapons black market, this node of characters connects to others based in Mexico. Menardo, a mestizo who amasses a fortune insuring wealthy Mexicans against "acts of God, mutinies, war, and revolution," is the central character who opposes revolution (260–61). He has a number of antagonistic connections to the revolution percolating in southern Mexico: his chauffeur, Tacho, feeds information about his boss' plans to an underground group led by Tacho's twin; and his wife, Alegría, has an affair with a leader in a Cuban Marxist group. *Almanac* concludes with the possibility that the American and Mexican resistance movements will merge to fight the war for territorial sovereignty and social justice.

As Joni Adamson, Shari Huhndorf, and Claudia Sadowski-Smith argue in their indispensable scholarship on *Almanac*, Silko's novel contributes a transnational scope to Native American studies and ecocritical studies alike. With a hemispheric imaginary, it "rewrites the history of the Americas from a transnational perspective that unites imperialism, slavery, and class struggle in a single story of ongoing land conflicts, and it attempts to negotiate a collective revolutionary identity based on histories shared by Native peoples across cultural and national boundaries."[10] Adamson focuses on the environmentalist dimensions of *Almanac*'s project "to expand the definitions of 'indigeneity'" and details the novel's vision of a transnational, coalitional environmental justice movement.[11] This scholarship has cemented *Almanac*'s place within the fields of American, Native American, and environmental studies. What interests me is how Silko's contribution to transnational and environmental imaginaries hinges on ecosickness, and yet, in the novel, ecosickness is a biotechnological phenomenon whose affective signature, anxiety, impinges on the very possibility of revolutionary action.

Sickness and its sometimes poisonous cures enter *Almanac*'s manifold plots: from Lecha's Demerol abuse and cancer that appear early in the novel to the International Holistic Healers Convention that concludes it, from environments marred by "poison smog . . . and the choking clouds that swirled off sewage treatment leaching fields" to visions of the Laguna plains "when the rain clouds had been plentiful and the grass and wildflowers had been belly high on the buffalo" (313, 758). Literary sickness often implies a narrative arc that travels from a negative condition of social, physical, or environmental dysfunction to redemption through healing and restored function. At moments, *Almanac* appears to fulfill this convention neatly. Channeling Yoeme, her grandmother, Lecha imagines that natural regeneration will replenish a degraded present, even if at the expense of humans themselves: "Old Yoeme had always said the earth would go on, the earth would outlast anything man did to it, including the atomic bomb. . . . Humans might not survive. The humans would not be a great loss to the earth. . . . Out of the dust grew the plants; the plants were consumed and became muscle and bone; and all the time, the energy had only been changing form, nothing had been lost or destroyed" (718–19). Through Yoeme's prophecies, Lecha envisions a form of transubstantiation: transferring land to its caretakers will first lead to agricultural regeneration and then to the growth of a new species of being.[12] Even if the coming conflagration annihilates the humans peopling the earth now, "the Earth is inviolate."[13] This prophecy, which circulates through tribes across the Americas, depends on a global sociopolitical revolution that will destroy all that the Destroyers have wrought in their turn.

With the lands restored and healed, Native peoples can again live the connection to the land that has been, if not severed, radically attenuated under white colonial oppression. In Yoeme's and others' voices, *Almanac* relates the place-based ethic that provides a model for U.S. environmentalisms premised on localism. As Ursula K. Heise clarifies, "the rhetoric of place in U.S. environmentalism . . . encompasses a whole range of sociocultural projects."[14] These projects include the kind of agrarianism Piercy's Mattapoisettans practice, bioregionalist education initiatives, and environmental justice groups' appeals to place experience. Common to these otherwise varied projects is faith that environmental accountability requires "spatial closeness."[15] We care best for what is near, a dictum that an author like Jan Zita Grover would readily endorse. Of interest to me in this chapter is that the ideals of human-environment connectedness often shade into body-land *merging* in environmental discourse. That is, "spatial closeness" is tantamount to oneness, to fusing with the more-than-human.[16] Merging is at once a trope and an experiential claim in environmental writing of disparate genres. One need look only to Annie Dillard's *The Pilgrim at Tinker Creek* (1974), Loren Eiseley's *The Immense Journey* (1957), or Gary Snyder's "second shaman song" (1960).

In many environmentalist texts, a simplified version of indigenous belief energizes and validates the rhetoric of merging. Paula Gunn Allen, a Laguna, Sioux, and Lebanese scholar, speaks a credo that provides a model for the identification of body and land in Anglo-American environmental writing. "We are the land," she declares. "To the best of my understanding, that is the fundamental idea that permeates American Indian life; the land (Mother) and the people (mothers) are the same."[17] Much Indian fiction narrates this deep connectedness through stories of illness and healing. If "we are the land," then an individual's sickness can stand in for the person's or the tribe's disrupted ties to place. This accounts for the medical imaginary through which Silko's first two novels depict the rift between humans and the more-than-human world. Indeed, if in *Ceremony* Tayo's "sickness was only part of something larger, and his cure would be found only in something great and inclusive of everything," in *Almanac* sickness has *become* "something great and inclusive of everything."[18] That is, while the environmental imagination of both texts centers on antonymic states of sickness and health, the balance tips to the former in *Almanac* while the telos of *Ceremony* is always the latter.

Almanac explores the status of the rhetoric of merging in a biotechnological age when we can no longer see soma and earth as protected domains of the natural. The novel still draws on affirmative Native models of the equivalence of human and environment, but, as technoscience increasingly intervenes in body and land to perpetuate injustice, the challenges to restoration are greater. Before arguing for how *Almanac* disrupts the narrative of planetary merging, I elaborate the conditions under which affective, cognitive, and spiritual immersion in place occur in the novel.[19] In *Almanac* people of many races and tribes make up the cast of revolutionaries. However, while racial and ethnic positioning do not predict for revolutionary fervor, it does predict for whether body-land merging is possible. While critics convincingly argue that *Almanac* does not have an essentialist idea of Indianness, ethnic and racial identity do condition the forms that environmental relation takes in the novel.[20] On the one hand, the novel naturalizes Native peoples' symbiotic relationship to their lands as both warrant and promise of territorial sovereignty. On the other hand, in order for nonindigenous peoples to connect to and even fuse with the nonhuman, they must face harm or death. Self-sacrifice is the pathway to oneness for this population. I argue that, once the text introduces self-injury as the pathway to ecological relatedness, violence taints the possibility of body-land merging.

A long history of inhabitation justifies the first conception of merging that applies to Indians. The territorial sovereignty that various parties seek both confirms and continues their identification with place. El Feo, an indigenous Mexican leading the rebellion south of the border, bases rebels' commitment to the revolution on whether they believe that the "ancestors had lived on the land for twenty thousand years continuously" (524). The long-standing, if now interrupted, habitation of the land forges a bond between past and present Native peoples and their homelands, a bond that is the precondition for any future sovereignty. This historical precedent warrants the struggle to "retake ancestral land all over the world. This was what earth's spirits wanted: her indigenous children who loved her and did not harm her" (712). A revolution is necessary to reinstate the lived relationship of love and care that will inaugurate ecological and tribal healing. Thus *Almanac* naturalizes the causal link between restoration of tribal sovereignty and planetary health, and its environmental imagination is intimately bound to its vision of social and political emancipation.[21] Silko builds this argument through a metonymic chain that links lands, tribes, and individual bodies. Old Yoeme's prophecy that dust will generate plants that produce "muscle and bone" is thus a partial synecdoche for *Almanac* as a whole (719): the healing of local ecosystems mends the Native person's body and psyche and stitches together a severed community.

The novel's final chapter, entitled "Home," invites us to read the text as attaining redemption and healing.[22] In "Home" the otherwise centripetal narrative comes back to a single point: Sterling and the Laguna Pueblo reservation to which he returns. Before diegetic time, the Tribal Council ostracized him for granting a Hollywood film crew access to a sacred stone snake that the filmmakers then stole. Where before Sterling had expressed his apostasy toward Laguna beliefs—"talk about religion or spirits had meant nothing to Sterling" (761)—he now awakens to the truth of his ancestors' prophecies when he sees signs that the earth and its people might heal and thrive. Buffalo have returned to the agriculturally devastated Great Plains "just as the Lakota and other Plains medicine people had prophesied" (758). The return of the stone snake seals these portents: "The snake was so near the [uranium mine] tailings it appeared as if it might be fleeing the mountains of wastes. This had led to rumors that the snake's message said the mine and all those who had made the mine had won.... But the following year uranium prices had plunged, and the mine had closed before it could devour the basalt mesa" (762).[23] Metaphoric relations between local conditions and global transformations supplement the metonymic chain linking environmental, communal, and individual renewal that runs through *Almanac*. The buffalo are analogues for the revolutionaries: the animals reclaim the Plains just as the rebels plan to recover the Americas. And the collapse of the uranium market stands in for the crumbling of U.S. neocolonialism. As Sterling notices these regenerative events, he is once again embedded in his ancestral ecosystem and the stories that infuse it. The chapter solidifies the interdependence of territorial restoration, community identity, and individual health when Sterling begins to abstain from drinking and recovers from his depression (757). All of these events signal the return to balance that many critics read in *Almanac*'s concluding chapter.[24]

As the end of Sterling's story coincides with the end of *Almanac*, it is tempting to assume that the healing of person, land, and tribe proposed in "Home" spreads to all but the Destroyers. However, indigeneity in fact determines whether salubrious environmental communion is possible. For those without a tribal affiliation, merging requires self-annihilation. It's as if the Destroyer race can do naught but destroy, even if the target is the self.[25] In a scene that brings diverse populations together, characters' discrepant access to body-land fusion becomes evident. The book's final part, "Prophecy," creates a crucible for *Almanac*'s cast. The occasion is the International Holistic Healers Convention that has been "called by natural and indigenous healers to discuss the earth's crisis" (718). This event and the novel's overarching position on pantribal unity allude to the Intercontinental Gathering of Indigenous People in the Americas, which met in Quito, Ecuador in 1990 to preempt celebrations of the quincentennial of Columbus's arrival in the Americas. One hundred and twenty Indian nations and organizations collaborated on the Quito Declaration to "ratify our resolute political project of self-determination and conquest of our autonomy" and "affirm our decision to defend our culture, education, and religion, . . . reclaiming and maintaining our own forms of spiritual life and communal coexistence, in an Intimate relationship with our Mother Earth."[26] When *Almanac* refracts this event as the Healers' Convention, earnest forms of environmental and social justice activism take place alongside New Age commodification of Native and Far Eastern spirituality and healing practices.[27]

The charlatans' opportunistic claims to cosmic connectedness contrast the depiction of human-nonhuman merging that ecological martyrs achieve in this same scene. The narrative introduces one such group, Green Vengeance, through their arresting footage of the bombing of Glen Canyon Dam. In the film "tiny figures dangled off ropes down the side of the dam.... The camera zoomed in for close-ups of each of the six eco-warriors.... Zeta had been thinking the six resembled spiders on a vast concrete wall when suddenly the giant video screen itself appeared to crack and shatter in slow motion, and the six spiderlike figures had disappeared in a white flash of smoke and dust. The entire top half of the dam structure had folded over, collapsing behind a giant wall of reddish water" (727). To contrast the frivolity of New Age commercialism, the narrator relates the audience's shock and presents the eco-warriors' martyrdom without irony. A fictional act that first appears in Edward Abbey's *The Monkey Wrench Gang* (1975), "dam-cracking" has become an activist desideratum by the time Silko pens her novel in the 1980s. Another source of inspiration is Earth First! which expressed its mission "to free our shackled rivers and restore the land" by simulating the dam's destruction.[28] In 1981 activists rolled a plastic "crack" down the façade of Glen Canyon Dam. Alluding to both the fictional and activist performances of environmental sabotage, *Almanac* raises the stakes of environmentalist resistance. To protest federal projects that fuel Western development and shatter ecological and tribal systems, Green Vengeance activists become suicide bombers and add gravity to an activist form that, in Abbey, was a humorous caper.

Comparing the activists to spiders, *Almanac* does not only establish the dam's humbling scale; the metaphor also suggests that the activists have let go of anthropocentrism. The suicide note they leave behind confirms that the bombing is both a blow to the infrastructure of Western growth and a strategy for planetary fusion. "'Rejoice! Mountains and valleys!'" the note proclaims. "'The mighty river runs free once more! Rejoice! We are no longer solitary beings alone and cut off. Now we are one with the earth, our mother; we are at one with the river'" (733). The Barefoot Hopi, a prophet, glosses the environmentalists' act: "the brave eco-warriors focused all the energy of their beings to set free the river, and so they merged instantly in the explosion of water and concrete and sandstone. They are no longer solitary human souls; they are part of a single configuration of energy'" (733). For the Green Vengeance warriors, planetary fusion corrects human alienation from the more-than-human. Crucially, however, fusion comes at the cost of existence itself for the non-Native characters.

In Silko's imagination, the realization of human and environmental oneness is possible. For Indian people, it occurs when one remembers the stories that establish an unmediated connection to homelands. For the nonindigenous, merging requires destruction, not of others' lives but of one's own. *Almanac* thus charts parallel if distinct avenues for body-earth fusion as counter to ecological and social colonization. While both direct-action environmentalism and tribal belonging can result in merging, the distinction between the paths to it is crucial, since violence taints the promise of ecstatic fusion in the former case. Once the text establishes self-destruction as the precondition for the Green Vengeance environmentalists' ecocentrism, body-land merging is no longer just a metonym for healing; it's also a resource for harm.

A scene blending the intensities of the pastoral and the thriller demonstrates the violence that the merging trope carries with it. *Almanac* flashes back to Lecha's past, when she used her clairvoyance to help police departments around the country track down killers and their victims' corpses.[29] In a Demerol haze, Lecha recalls a haunting case in San Diego. The sound of the ocean channels her into a serial killer's consciousness where she sees the murderer's rituals and the perverse logic of care that motivates them.

> The eyes are gone. The sand fills the sockets. Now the boy has eyes the color of sand . . .
>
> He imagines the boys are trees that he must go tend from time to time. He uncovers them tenderly. To see how they are developing. They thrive best at the foot of the big dunes. Out in the flats they can't take root. . . .
>
> He realizes they are trees while he is touching them. He fondles the boys between their legs, and a branch sprouts and pushes out. The tips are soft leaf-bud moist with sap. He never means to squeeze too hard or to crush. But they are tender, fragile. He plants carefully and prays for tall trees. He dreams of towering oaks and spruce that lean and sway but do not break in summer storms. (140–41)

The killer cloaks his violence in pastoral tropes, and these tropes in turn connote psychological sickness for the reader. The killer rhetorically turns the boys into a plot of land that demands his assiduous attention, a form of care that obscures the acts of violation and murder. The dead boys merge with their ocean surroundings even as the murderer turns them into trees that the dunes cannot sustain. "Growth" is the term on which the elision of sexual violence and gardening depends. Having metaphorized sexual tumescence through plant growth, the killer handles the corpses as if they were plants and he the gardener who helps them flourish. Forms of the word *tend* appear three times in these lines and code the killer's injurious touch as touching care for body and land. Maggie Ann Bowers is right to hear in *Almanac* "recognition of the need for a close personal relationship to a specific landscape," but this early incident adds a macabre dimension to the merging of body and land.[30] In the story of perverse pastoralism, humans become more-than-human not as a check on cruelty and exploitation but as a vehicle for them.

Undoubtedly, *Almanac of the Dead* shares with forms of U.S. environmentalism a confidence that closeness to nature fosters environmental responsibility, especially when this closeness intensifies to fusion with the more-than-human. However, the eco-warriors' suicide, while still lauded, introduces into the novel an environmentality tinged with violence and negative affect that the episode of the gardening killer only heightens. *Almanac* is of two minds. On the one hand, Native characters' expressions of belonging to place show environmental connection to be indispensable to universal healing, a position that's in line with literary and metaphysical traditions. Yet, on the other hand, the novel's representation of environmental relation introduces a violence for which care is meant to be a cure. Sickness has so overtaken the Americas that merging with the land may promote healing according to Native and environmentalist templates, but it may just as well mask injury and destruction. As the rest of this chapter argues, *Almanac*'s anxious biotechnological plots also transform earth and soma into conflictual domains and thereby jam the narrative of environmental and somatic restoration. The novel's technoscientific imaginary thus unites body and land on other terms: as generators of anxiety from which Silko, unlike Piercy, does not provide release.

INIQUITOUS INTERVENTIONS

Just as *Almanac of the Dead* promotes the therapeutic narrative of body-land connectedness even as it disrupts it, the novel alludes to an optimistic narrative of technological progress even as it undermines it. With its dual settings, *Woman on the Edge of Time* demonstrates that biotechnologies can sometimes alleviate biophysical, environmental, and social sickness. While Piercy's novel proposes that genetic modification can improve labor conditions and promote equality, *Almanac* shows that somatic and ecological interventions are inherently iniquitous. Technologies such as artificial biospheres, regenerative medicine, and genetic selection violate the boundaries between the given and the engineered, between the enduring and the new, and between the healthy and the contaminated. They destabilize the categories of body and earth and thereby destabilize the merging of these domains that heralds planetary restoration.

Almanac confirms that, with the advent of genetic modification in the 1970s, we began to move "from an age concerned with *representing* organisms and their processes—an age concerned with discovery—into a technological age, one concerned with *intervention*, whose telos is that of rewriting and transforming life."[31] For Nikolas Rose, enhanced ability "to control, manage, engineer, reshape and modulate the very vital capacities of human beings as living creatures" characterizes this new interventionist age.[32] Writing when the biotechnological age is just commencing, Piercy imagines the horrific implications of *manage* and *engineer*. *Woman on the Edge of Time* narrates the myriad ways the bodies and minds of the poor and people of color are subjected to medical surveillance and control. As the ability to manipulate human and nonhuman life expands in Silko's book, dilemmas of management, choice, and responsibility challenge ethical models that are predicated on the strict separation of life and technology. If these domains can interpenetrate beneficially in the Mattapoisett sections of Piercy's novel, they no longer can when Silko writes a decade later. While for Rose the "technological age" empowers individuals to create enabling biological identities and pursue new affinities based on those identities, Silko is not so sanguine. Her text aligns technoscience with inequalities based on ethnic, racial, and class identity. A speculative fiction that straddles the mimetic and the predictive, *Almanac* amplifies scenarios of technological innovation and portrays body and land as zones of anxiety.[33] The proliferation of biotechnology thus generates negative affect in *Almanac* not only because it disrupts the script of body-land merging on which the novel's revolutionary promise depends but also because these technologies aid Destroyer efforts to maintain privilege.

In *Almanac* almost cartoonishly evil characters promote biotechnological projects that preserve the value of genetic "purity" at the expense of the poor. The Spaniard Serlo, a "genuine blue blood," turns his estate in Argentina into a fortified research institute and "stronghold for those of *sangre pura* as unrest and revolutions continued to sweep through" Latin America and the U.S. (535, 541). With Serlo, *Almanac* intertwines genetic and environmental technologies under the banner of eugenics. As intensifying environmental catastrophes, race riots, and economic upheaval threaten the ruling class, Serlo commissions a Swiss scientist to design "Alternative Earth modules" to which the rich can escape while the poor destroy themselves. Serlo collects and preserves "pure" resources for these units, an effort that parallels his preservation of unsoiled bloodlines through eugenics research. "In the end," Serlo reasons, "the earth would be uninhabitable. The Alternative Earth modules would be loaded with the last of the earth's uncontaminated soil, water, and oxygen and would be launched by immense rockets into high orbits around the earth.... The select few would continue as they always had, gliding in luxury and ease across polished decks of steel and glass islands where they looked down on earth" (542). The modules are not so much survival bunkers as luxury cruise liners that offer neo-"colonists" an insulated, aerial vantage on ecological and social disaster. As an escapist adaptation to endangerment, Serlo's closed ecosystems exploit fears about diminishing resources and threats to elites' continued domination of the poor.

Rhetorically, Serlo's eugenics and biosphere projects share origins in the threat of contamination. As Adamson argues, the Alternative

Earth Units parody the rhetoric of wilderness preservation, in which "some species are viewed as 'contaminants' and targeted for removal, but other species are viewed as 'endangered' and targeted for protection."[34] The "species" concept extends to humans as well, some of whom—the poor and people of color—are sacrificed to ideals of "'natural settings'" and racial purity (728). Genetic and alternative ecosystem research also relate causally because ecological destruction threatens racial purity in Serlo's world view. His genetic engineering and artificial insemination projects are futile on their own because "there was little use in bringing a genetically superior man into a world crowded and polluted by the degenerate masses" (546). Because pure blood cannot flourish in impure earth, Serlo frames degeneracy in terms with strong environmental resonances. Focalizing Serlo, the narrator anticipates the social and economic Armageddon from which the wealthy must protect themselves: "Epidemics, accompanied by famine, had triggered unrest. Mass migrations to the North, to the U.S. border, by starving Indians had already begun in Mexico. Serlo and the others with the 'hidden agenda' had only a few more years to prepare before the world was lost to chaos. Brown people would inherit the earth like the cockroaches unless Serlo and the others were successful at the institute" (561). The portents of apocalypse intensify this passage's already extreme rhetoric. In addition to disease, starvation, and human exoduses, the reference to cockroaches signals that the end times are nigh. Standing in here for Mexican and Indian "invasion," the vile insects anticipate ecological and social collapse in racist terms. Resilient immigrants from the south will infest their new northern habitat and eventually crowd out the "noble" wealthy whites (561). To Serlo's racist mind, as undesirable "brown" bodies contaminate and exhaust land, air, and water, preserving blood and earth become exigent, interdependent projects. Through this greenwashed eugenicism, the novel figures a wicked environmental consciousness in which ecological reconstruction paves the way to racial preservation.

Serlo's venture is an obvious fictionalization of the real-life Biosphere 2 initiative that was located in Oracle, Arizona, only twenty miles northeast of Silko's home of Tucson.[35] With this allusion, *Almanac* calls out the racist overtones of some contemporary U.S.

environmentalisms. In 1984 ecologist and engineer John Allen and Texas oil tycoon Ed Bass developed the concept for Biosphere 2 out of concern for the "increasing population and the scale of human activity" that concerned Allen in the 1960s.[36] Biosphere 2 was designed as "a closed system with potential as a pattern for future off-world colonies, a prototypical tool for studying closed systems and a test platform for technology."[37] The geodesic terrarium replicates five ecosystems (rainforest, fog desert, savannah, marsh, and ocean) using soil, plants, and water trucked in from around the country, and it originally sustained eight "biospherians" entirely sealed off from the outside world for two years.[38] Though designed by scientists of varying disciplinary backgrounds, Biosphere 2 lacked credibility from the outset. Certainly many early reports lauded the project's aims to explore the complex interconnectedness of human life and ecosystemic processes,[39] but skeptical analyses were forthcoming. Allen and company's ambitions were outsized, and critics portended the failures that did indeed come. Moreover, though Biosphere 2 was marketed as an "environmentally sensitive community" and an experiment that would contribute to ecological knowledge, journalists pointed to plans to build "RV parks, shopping centers, gas stations . . . and—yes—a golf course" in the neighboring town of Oracle as evidence that it was in fact meant to "cash in on the current environmental craze and turn a scientifically anemic experiment into an amusement park gold mine."[40]

Almanac's dark parody of the endeavor located in Silko's backyard hinges on racial and class tensions absent from the journalistic accounts of Biosphere 2. None of the venture's early critics mentions the disturbing irony of having an all-white crew cultivating "uncontaminated" soil in the Mexico-U.S. borderlands, a region suffused with tensions over immigrant "invasion." Through Serlo's eugenicism, the novel casts closed ecosystems research as a perversion of environmental action. The novel brings the racist implications of Allen's project to the fore as it exposes the iniquitous aspects of so-called technoscientific progress. In its wide-ranging conversation with American environmentalism, *Almanac* refracts skepticism about the environmentalist bona fides of emergent technologies. But it avoids dismissive skepticism or Luddism in favor of urgent anxiety about deep-seated inequities that motivate biotech and that biotech in turn promotes.

A concern that environmentalism could be hijacked for ethnic purification and for the continued dispossession of nonwhites and the poor threads through *Almanac*. In particular, it calls out the rhetoric of pollution that energizes eugenicist and ecocentric ideologies alike. While the novel undoubtedly depicts coalitions for environmental justice forming across tribes, nations, and races, it also shows the ideological rifts that can cloud this ideal.[41] Most importantly for my purposes here, advocates of social justice distrust deep ecologists for spouting another of "the white man's lies" (406). Clinton, the black leader of the "Army of the Homeless," propagates this thought through his incendiary radio program. For him, white environmentalism cannot be a pathway to liberation and equality.

> Clinton did not trust the so-called "defenders of Planet Earth." Something about their choice of words had made Clinton uneasy. Clinton was suspicious whenever he heard the word *pollution*. Human beings had been exterminated strictly for "health" purposes by Europeans too often. Lately Clinton had seen ads purchased by so-called "deep ecologists." The ads blamed earth's pollution not on industrial wastes— hydrocarbons and radiation—but on overpopulation. It was no coincidence the Green Party originated in Germany. "Too many people" meant "too many *brown-skinned* people." . . . "Deep ecologists" invariably ended their magazine ads with "Stop immigration!" and "Close the borders!" Clinton had to chuckle. The Europeans had managed to dirty up the good land and good water around the world in less than five hundred years. Now the despoilers wanted the last bits of living earth for themselves alone. (415)

Strategically positioned before the exposition of Serlo's eugenicist research, Clinton's theory calls out the overt and latent racism present in particular strains of environmental discourse.[42] The repeated scare quotes announce his deep suspicion toward environmentalisms that do not only privilege ecological protection over the rights of the marginalized but also attempt to eliminate the marginalized entirely.

Referring to Germany (read Nazism), the passage suggests that deep ecologists who advocate a nonhierarchic system of species value are racists at heart and would endorse Serlo's ecotechnological project out of a shared populationism.[43] For Clinton, it is a small step from the deep ecological principle that "the flourishing of nonhuman life is compatible with a substantial decrease of the human population" to the idea that "decrease" requires annihilating or displacing polluting "*brown-skinned* people."[44] *Pollution* is thus a polluted term with both environmental and sociocultural resonances. Just as construction of factories, highways, power plants, and landfills differentially affects people of color and the poor, efforts to eradicate pollution cover up undercurrents of racism. At the same time as *Almanac*'s narrative of restoration appeals to the inclusionary model of environmental justice, it also delineates "the greening of hate" that can be a barrier to the coalitional politics that it urges.[45]

Clinton advances a line of argument that Ramachandra Guha forwards when he worries about the consequences of deep ecology's radical biocentrism. Because this philosophical disposition "has begun to act as a check on man's arrogance and ecological hubris," it is "to be welcomed."[46] However, because deep ecology privileges the preservation of so-called wild spaces—that is, those on which indigenous and poor peoples have long made their living—it perpetuates an imperialist agenda. In Guha's words, deep ecology "provides, perhaps unwittingly, a justification for the continuation of . . . narrow and inequitable conservation practices" and, just as crucially, neglects the environmental "inequalities *within* human society" such as racist urban development patterns and uneven access to food and land.[47]

Clinton echoes Guha's points and adds that population debates in particular are a locus for the conflict between social justice and preservation of pristine spaces. He reads into "European" campaigns against pollution a chauvinistic desire to ensure the physical longevity of the upper classes. On this point, *Almanac* shifts focus from exclusionary ecotechnologies to biomedical technologies that reinforce somatic privilege along class lines. Just as Serlo's Alternative Earth units are instruments of hate and inequality, biomedical technologies injure the poor. Both sets of innovations exploit existing socioeconomic and racial disparities in order to enhance them. Regenerative medicine and trade in human blood, tissue, and organs constitute the novel's "biomedical imaginary," which Catherine Waldby defines as "the speculative, propositional fabric of medical thought,... a kind of speculative thought which supplements the more strictly systematic, properly scientific, thought of medicine, its deductive strategies and empirical epistemologies."[48] Waldby refers here to the wish-fulfilling narratives that drive biomedical research within the laboratory. Novels such as *Almanac* and *Woman on the Edge of Time* demonstrate that speculative fiction can reveal the fiction making of research agendas, that is, how they are fueled by "disavowed dream work."[49] In Silko's text the desire for profit certainly drives biotechnological projects, but, more importantly, the desire to cement white privilege at the expense of the marginalized motivates these schemes.[50]

Secrecy and the entanglement of innovation and personal desire abet racist technoscience in *Almanac*. The novel gives voice to these drives through Eddie Trigg, whose "dream work" yokes together land development, traffic in human bioproducts, and regenerative medicine. Like Serlo's research center, which operates "in complete seclusion" (542), Trigg's ventures are private in three senses: they are privately funded, unregulated, and deeply personal. His entrepreneurial ambitions stem from the trauma of becoming paraplegic in an alcohol-fueled car accident in college.[51] Trigg's motivations and plans come to light when his lover, Leah Blue, steals his therapeutic diaries. First, he intends to manipulate Tucson's real estate market and use the profits to pay for research into genetic therapies designed to restore function to his body. "Trigg bragged to Leah that blood-plasma donation centers busted neighborhoods and drove property prices down without moving in blacks or Mexicans. With property prices down, Trigg came and cleaned up, buying most property for forty cents on the dollar" (379). He sees that the clientele for blood donation centers are typically the poor, people of color, and drug users and rigs real estate values by moving these so-called undesirable populations into the neighborhood. With prices low, he then sweeps back in, closes the blood centers, and gentrifies the area. In the plot involving Trigg and Leah Blue, entrepreneurs manipulate disparities between the impoverished and the wealthy and between the sick and the well in order to profit from land and biomedical ventures. Inspired by Leah, Trigg expands to other domains: from plasma centers, he trades in "organs and other valuable human tissue" (387). In addition to supplying organs to those needing transplants, the lovers will supply housing and the necessary medical infrastructure: "Leah had got the idea for a kidney dialysis machine that would serve the sector of town houses and condominiums that would presumably be bought by kidney patients and their families or by health insurers to house their Arizona dialysis patients" (387). Just as one cannot lose when investing in Arizona's booming real estate market, the "eventual miracle of medical science and high technology" is certain to return dividends to those who exploit the desperate and leverage health for profit (380). With the capital from the medical "miracle" of organ transplantation, Trigg will invest in another "breakthrough technology.... The rewiring of human nerves severed or badly crushed" (386).

Of course seeking a cure for paraplegia is not dubious on its own. But Silko makes it so when her character participates in what Rachel Stein calls "the biotechnological colonization of the vulnerable."[52] That is, Trigg treats the bodies of the urban poor as so many "usable, extractable, tradable" parts for his entrepreneurial machine.[53] As the text explains how Trigg sources the "biomaterials" that his researchers need to conduct their experiments in regenerative medicine (389), evidence mounts that, as Ann Stanford writes, "the lure of the infinite possibilities promised by technology is coupled with a widespread disregard for all human life that does not directly serve the interests of the wealthy."[54] "Disregard" is too tepid a description of how Trigg acquires biomaterials, since he does so through sexualized murder. Drunk and high, he confesses to an employee that he enticed a homeless hitchhiker to his blood center and "gave him a blow job while his blood filled pint bags; the victim relaxed in the chair with his eyes closed, unaware he was being murdered.... Trigg blames the homeless men. Trigg blames them for being easy prey.... They got a favor from him. To go out taking head from him.... They were human debris. Human refuse. Only a few had organs of sufficient quality for transplant use" (444). Trigg inverts a sexual hierarchy while reinstating a socioeconomic and racial one. His donor-victims receive Trigg's "favor" of sexual pleasure, yet Trigg is simultaneously sucking out their lifeblood. Whereas, in the texts Stein analyzes, women's bodies are vulnerable to "colonization," in *Almanac* indigent people of color are the primary targets. Trigg justifies his actions through his class privilege and his belief that "nobody ever notices when [the homeless] are gone" (444). They are disposable, or, more accurately, they are only organic raw materials. The poor, blacks, and Mexicans have no value until they become the components in a research agenda whose goal is Trigg's personal well-being. Trigg's actions highlight precisely the distortion of "biovalue" that concerns Waldby in her critique of how biomedicine treats the body as data or resource. She defines biovalue as "a surplus value of vitality and instrumental knowledge which can be placed at the disposal of the human subject. This surplus value is produced through setting up certain kinds of hierarchies in which marginal forms of vitality—the foetal, the cadaverous and extracted tissue, as well as the bodies and body parts of the *socially* marginal—are transformed into technologies to aid in the intensification of vitality for other living beings."[55] Having placed himself at the top of a social and racial hierarchy, one that Serlo also enforces, Trigg assumes ownership of the biovalue of "the *socially* marginal." He fragments and instrumentalizes marginalized bodies, treating them as mere ingredients for his "breakthrough" rather than as whole humans in their own right.

Serlo's and Trigg's ventures express Paul Rabinow and Nikolas Rose's understanding of "biopower" as "fundamentally dependent on the domination, exploitation, expropriation and, in some cases, elimination of the vital existence of some or all subjects over whom it is exercised, . . . a form of power which ultimately rests on the power of some to threaten the death of others."[56] Biopower in *Almanac* is a race- and class-based privilege that capitalizes on the increasing malleability and vulnerability of life itself and on the commodification of bodily materials. Here malleability is a potential boon to those such as Trigg and Serlo who can fund projects that serve their ideological ambitions and decrease their own vulnerability. For the indigent and "*brown-skinned* people," on the other hand, the malleability of life only increases vulnerability to injury and even death (415). Silko refracts late twentieth-century technological interventions that reconfigure vitality to show that biologically generated capital only flows to those with a racially and economically marked form of agency.[57]

The fragmentation of body and land is the pivot on which *Almanac*'s imagination of ecosickness turns. Just as space can be parceled out and manipulated to serve colonialist and capitalist desires, the body can be dismantled and resourced for its economic and scientific value. A modular conception of the land underpins Serlo's Alternative Earth project; the whole is less than the sum of its parts, which can be artificially replicated and recombined. Leah shares this perspective on the land. She treats it as a slate that can be wiped clean and reconstituted according to her will. In this respect, "it is realty that both consumes and subsumes reality," as Rebecca Tillett notes.[58] Leah smiles on the "huge tracts of desert that had been bulldozed into gridworks scraped clean of cactus" and dreams that "the city limits of Tucson and surrounding Pima County were a gridwork of colored squares for Chinese checkers" (359, 360). The desert is a board for a rationalized game of economic and environmental exploitation. Serlo's eugenics research and Trigg's biotechnology schemes rest on a similarly modular conception of the body. They substantiate Ian Hacking's thesis that neo-Cartesianism again dominates conceptions of the body in the late twentieth and early twenty-first centuries. Once, organs and genes "constituted different aspects of our bodily essence," Hacking contends. "In different ways they were inseparable from our selves. We could not do much about them. Now we can."[59] Consequently, "our bodies are likely to become more other."[60] *Almanac* establishes that, the more they're integrated into ideological and capitalistic projects such as Trigg's, bodies and environments become increasingly dis-integrated from the wholes they once constituted.

Almanac of the Dead heightens readerly anxiety about expanding scientific innovation by depicting actual emergent eco- and biotechnologies, but rooting them in racism and somatic and environmental exploitation. It is a forerunner to twenty-first-century novels such as Atwood's *Oryx and Crake* and Paolo Bacigalupi's *The Windup Girl* (2009) that push environmental thinkers to account for whether biotechnological developments can fit into an equitable future. Where *Woman on the Edge of Time* envisions genetic technologies as integral to a salubrious relation to the land and a just community, *Almanac* envisions the technologies of life as fragmenting the "single configuration" of body and land on which universal healing depends (733). Ultimately, Silko's novel suggests that there are no conditions under which we should intervene in life itself. If, as Hacking asserts, "new conceptions and practices . . . are assimilated to core emotions connected with living and dying, health, well-being, and distress," then anxiety is the "core emotion" that Silko connects to pervasive ecological and bodily sickness as well as the technological interventions that exacerbate rather than cure it.[61] It might not surprise us that the affect of anxiety circulates in narratives of ecosickness, as it does in doomsday environmentalism and technophobia in general. Yet, however common anxiety is to these discourses, *Almanac* is peculiar in how it reveals the inefficacy of the emotion that permeates its critique. The text suggests that anxiety is not always an effective catalytic for revolution and restoration. As the remainder of this chapter shows, the very conventions that tag the novel as apocalyptic minimize anxiety's revolutionary efficacy.

ANXIOUS APOCALYPSE

In demonizing technoscience through villainous characters like Serlo, Trigg, and Leah, *Almanac of the Dead* aligns itself with a technopessimism that runs through some strains of environmentalism and ecocriticism. The novel expresses profound skepticism, even hostility, toward technoscience as an environmental and somatic threat, even as it attempts, like many environmental writers and scholars, to bridge the divide between cultural studies and the sciences. According to Heise, technopessimism runs through environmental thought because "science is viewed as a root cause of environmental deterioration, both in that it has cast nature as an object to be analyzed and manipulated and in that it has provided the means of exploiting nature more radically than was possible by premodern means."[62] In Silko's book, regenerative medicine, eugenics research, and alternative ecosystems exemplify biotechnology's capacity to transform life itself and hence buttress the structural inequalities that cement Destroyer privilege.

Almanac explodes the narrative of technological progress. Biotechnologies aggravate rather than heal the social, environmental, and somatic ills that are the hallmarks of apocalypticism, one of the novel's dominant modes. Technoscience is at the center of two complementary strains of apocalyptic thought running through contemporary U.S. culture and Silko's text in particular: the secular and the religious. In religious apocalypticism, the abuses of technological prowess that *Almanac* documents would be read as signs of divine discontent with humans' misuse of their earthly inheritance. Michael Barkun argues that, no matter whether the portents of apocalypse are natural (in his example, earthquakes and famines) or social and political (sexual licentiousness and threats to the Jewish state in Israel), religious apocalypticism finds the source of catastrophe in the divine. This mode of crisis thinking shares space with a secular version that surges in the 1970s and 1980s. Secular apocalypticism in Barkun's account arises from "nuclear war, spiritual and ideological exhaustion, environmental degradation, and overpopulation and the depletion of basic resources."[63] With these crises as their concern, "the secular apocalypticists insist upon the interrelatedness of the human and natural worlds."[64] For these thinkers, as for *Almanac*, the interdependence of body and land symbolizes the restoration of individual, social, and environmental health, while the alienation from body and land stands in for global collapse. In the secular vision progress and modernization implode: the technological fruits of human ingenuity such as militarized science and industrial development are *causes* rather than *signs* of collapse. Barkun is quick to explain that secular apocalypticism does not supplant the long-standing tradition of religious millenarian thinking. Rather, the "growing willingness to contest the promise of technology" coincides with the "growth in prestige enjoyed by science" and coexists with religious ideas about the end times.[65] By the early 1990s, when Silko's novel hits bookshelves, biomedicine and ecotechnology join nuclear weaponry as fuels for techno-anxiety.[66]

Almanac commingles the religious and the secular to imagine a cataclysm centered on interventionist technologies that exacerbate ecosickness and injustice. Indeed, for all of the novel's narrative inventiveness, it cleaves quite closely to conventional apocalyptic imagery. Earthquakes, volcanic eruptions, droughts, tidal waves, and landslides: these natural disasters and endemic depravity situate *Almanac* in the apocalyptic tradition. A fragment from the Yaqui almanacs that Lecha is archiving introduces the novel's religious lexicon for apocalypse.[67] Just as the Bible's Book of Revelations warns of the calamities that will arrive on Judgment Day, the almanacs anticipate that "plague, earthquake, drought, famine, incest, insanity, war, and betrayal" will trail behind "Death-Eye Dog," the herald of an epoch of evil and global destruction (572).[68] While *Almanac* does not share a Christian view of the afterlife and, through Yoeme, even envisions a revitalized but possibly human-less earth (718), the text shares an idiom of apocalypse with Abrahamic religions. This idiom complements the novel's secular apocalypticism centered on iniquitous biotechnological interventions. Specifically, *Almanac*'s biotech plots transform the promise that human-land connectedness will lead to harmony and restoration into anxiety about whether the planet and human bodies will endure as such in an age of technoscience. Soma and earth become the repositories for the text's apocalypticism and the pervasive anxiety that it generates. In effect, the novel's anxiety about technological apocalypticism shatters the integrity and therapeutic merger of body and land. This raises the question whether *Almanac*'s apocalypticism and its plots of revolution and regeneration are compatible. Can anxiety be the affective impetus to revolutionary fervor and the restoration of body, community, and environment?

To approach this question, it's worth elaborating in more detail how the novel creates its apocalyptic feeling. It develops through the depiction of sick bodies and environments and the biotechnological interventions into them that signify collapse, features of the novel this chapter has already explored. *Almanac* also evinces a representational tendency that *Ecosickness* has been tracking: human bodies and the more-than-human realm increasingly interanimate metaphorically in contemporary fiction. That is to say, bodies assume their contours through environmental figures and space is medicalized. Silko's novel uses this representational strategy to medicalize environmental consciousness, as I've argued other ecosickness fictions do. But here the strategy also turns body and land into sites of violence and increases the sense of apocalyptic doom.[69] *Almanac*'s metaphorical habits augment its anxious apocalypticism at the same time as they formally mark the novel as an ecosickness fiction.

The southwest desert, a place long thought to be hostile to human settlement, is an apt setting for bio- and ecotechnological projects that announce the vulnerability of life itself. The author explains her fascination with the negative feelings that attach to her adopted town in "Notes on *Almanac of the Dead*." Researching the history of the Sonora desert region "from the time of the Portuguese monster de Guzman . . . to the present," she "began to wonder if there was something in the very bedrock, in the very depths of the earth beneath Tucson, that caused such treachery, such greed and cruelty."[70] For Silko, violence bubbles up from the place itself and produces a palpable unease. Against notions that the land is inherently salvific, this comment indicates that the Sonoran Desert does not merely suffer from the wickedness of colonialists and developers; the land in fact generates barbarism. For as much as Silko's desert home is a place to which she, like the Yaquis in *Almanac*, "felt loyalty," it provokes aversion due to the history of violence that it connotes and the threat that it presents (223). In representing the desert using a somatic metaphorics, *Almanac* evokes this feeling of physical endangerment.

This sense of threat does not prevent the U.S. Southwest from drawing significant economic and real estate investment in the 1980s, when Silko writes *Almanac*, and in the decades up to the global recession of 2008. In the novel Leah Blue represents the surge in development during the boom and the ecological manipulation that it requires. As she surveys the site for her new venture, Venice, Arizona, the future development materializes as a watery foil to the hostile desert:[71]

> She liked to step on the concrete pedestal at the deep wellhead, to gaze out over the brownish desert shrubs and grayish desert gravel and visualize the sleek, low villas of pale marble with red bougainvilleas and even water lilies for the floating gardens in the canals. She could not understand why the Indians or the environmentalists had bothered to sue even if her deep wells *did* harm other wells or natural springs . . . ; what possible good was this desert anyway? Full of poisonous snakes, sharp rocks, and cactus! Leah knew she was not alone in this feeling of repulsion; most people who saw the cactus and rocky hills for the first time agreed the desert was ugly. (750)[72]

The desert's dusty, drab, "brownish" and "grayish palette" contrasts the purity and vibrancy of the pale marble of the Mediterranean architecture and the red of the introduced plants. Devoid of pigment, the desert seems to be devoid of life, and this emptiness "mean[s] danger and death" (637). Leah's observations continue a tradition of denigrating the American desert that is as long as the history of white settlement in the region. In his study of topophilia, Yi-Fu Tuan offers the desert as a counterexample to pleasurable place feeling. He cites Lieutenant J. H. Simpson's assessment of New Mexico in 1849: it is "sickening-colored"; "until familiarity reconciles you to the sight," the desert cannot but produce "a sensation of loathing."[73]

Silko enhances the repulsion that typifies the desert by setting *Almanac* in a deep drought that, unlike the one that structures *Ceremony*, has no foreseeable solution. After a long absence from Tucson, Seese remarks that "even the so-called desert 'landscaping' was gaunt"; adaptation through xeriscaping cannot erase the signs of ecological—and, in the novel's logic, social and somatic—crisis (64). *Gaunt* is an adjective we associate with human description rather than landscape description, and Silko uses it to intensify the sense of ecological emergency that water-intensive developments like Venice create. This strategy continues as the narrative deploys human anatomical synecdoches to figure the apocalyptic effects of drought. "The prickly pear and cholla cactus had shriveled into leathery, green tongues. The ribs of the giant saguaros had shrunk into themselves. The date palms and short Mexican palms were sloughing scaly, gray fronds, many of which had broken in the high winds and lay scattered in the street" (64). Seese's perceptions medicalize environmental perception under the sign of crisis. "Green tongues," shrunken "ribs," and "sloughing" introduce a representational strategy borrowed from urban description, in which metaphors of human disease, often cancer, signify urban squalor and depravity.[74] In *Almanac* metaphors of the sick body imbricate human and environment, but also accentuate catastrophes that cross ecological, social, and physiological systems. Indeed, there's something grotesque and anxious about a figurative strategy that assigns a landscape human parts in order to communicate injury to the total environment.[75] A proximate passage uses a similar lexicon to transform bodies into desiccated plant life and forms a chiasmic pair with the medicalized landscape description. Seese recalls how her ex-lover David had "seen too much loose skin during those years" when he prostituted himself to gay millionaires in Malibu (59).[76] David compares his clients' penises to "withered vines and grapes shrunken like raisins; layers and layers of grayish crepelike skin dropping off" (59). *Withered, shrunken,* and *grayish* carry the drought from earth to body, from Arizona to California, and from the narrative present to a past outside of diegetic time.

When not painting the Southwest with a palette of drought, *Almanac* paints it in blood. In this respect, the novel echoes *Woman on the Edge of Time*, which describes sunrise as the break of a "day already bleeding at the edges" to inject the atmosphere with indeterminate menace (26). In Silko's text twilight is the occasion for this metaphor. Looking out over a range of New Mexican mountains, Leah's nephew observes that "the colors changed rapidly after the sun set. The sky ran in streams of ruby and burgundy, and the puffy clouds clotted the colors darker, . . . into the red of dried blood. . . . Then the light faded and the breeze slashed at the ricegrass and yuccas. The cooling brought with it deep blues and deep purple bruising the flanks of the low, sandy hills" (370–71). The narrative borrows from the repertoire of horror to depict a desert sunset, a scene typically idealized in landscape art. Sky, clouds, and hills assume the shape of a body in order to suffer physical damage. After the color red thickens through the process of clotting, it not only paints the hills but also transforms them into flanks susceptible to bruising. In these lines, blood is not only a metaphor; bleeding also guides the gaze. That is, as the character's eyes move across the scenery, the wind "slashes" the plants, suggesting that this is the source of the gore. Topography mutates into vulnerable anatomy both to amplify the mood of catastrophe that permeates the diegesis and to confirm that catastrophe will express itself through erstwhile domains of the natural: body and land or, in this case, land turned to blood.

In Grover's *North Enough* the metaphorical exchanges between the human and the nonhuman figure the tension between the surface damage of AIDS bodies and clear-cut forests and their alluring depths of beauty. The narrator of *Infinite Jest* apprehends the novel's settings in detailed anatomical terms, and this strategy of imbricating soma and earth creates an environmental consciousness of malleability and porosity. While in these other ecosickness fictions medicalizing space generates ameliorative forms of discord and disgust, in *Almanac* it creates a totalizing picture of catastrophe, of scorched lands and bodies bleeding into each other. The transfers between body and earth in *Almanac*'s metaphors lend the text an apocalyptic tone that produces anxiety. In other words, as body becomes earth and earth becomes body through the tropes I have detailed here, hope for repair dissipates and doom creeps in. If the "materiality of the body [is] the ultimate object of technologies of fear," as Brian Massumi claims, then rendering the environment through the body strategically activates the negative affects that suffuse *Almanac*.[77] In the novel body and land merge tropologically, but, because of the negative charge of these tropes, their fusion cannot simply connote therapeutic connectedness and restoration. Body and environment mutate into threatened and also threatening objects that constitute the text's apocalyptic atmosphere. For this reason and because of their vulnerability to biotechnological intervention, soma and earth trigger the anxiety of apocalypse.

With this texture of metaphors, Silko's ecosickness fiction borrows from the rhetoric of American literary apocalypticism, but also diverges from it to bring into focus the somatic and environmental crises of a biotech age. The novel certainly adapts Mayan, Yaqui, and other Native traditions of prophecy, as Adamson and other critics have pointed out.[78] However, I want to shift attention to two other traditions of apocalyptic rhetoric in the U.S.—environmental millenarianism and the jeremiad—to underscore Silko's generic eclecticism and to illuminate the kinds of critique her novel performs. Jimmie Killingsworth and Jacqueline Palmer delineate the first of these traditions in their study of "millennial ecology," "a thoroughgoing critique of progressive ideology [that] fosters a totalizing vision of political transformation."[79] Millennial environmental rhetoric places apocalypticism and social action on a continuum; it does not so much predict catastrophe as galvanize a movement. That is, in Killingsworth and Palmer's account, environmental thinkers—Rachel Carson, Paul Ehrlich, and Barry Commoner represent the trend in the 1960s—

foretell destruction to energize a sympathetic base. For many of these writers, the apocalypse is not inevitable, and millennial rhetoric is part of "a radical undertaking, the aim of which is to root out the conceptual underpinnings of a polluting culture—in particular the ideal of progress—and establish new foundations for communal action."[80] Especially with the revival of environmental apocalypticism in the 1990s, Killingsworth and Palmer demonstrate, writers mute their rhetoric precisely because action is their objective. Their doomsday scenarios "are not to be taken literally. Their aim is not to predict the future but to change it."[81] Apocalypticism and protest are compatible then only to the extent that this rhetorical mode remains just that, rhetorical, and that catastrophe is avoidable rather than a foregone conclusion. As a contemporary variant on millennial ecology, *Almanac*'s depiction of errant biotechnology guts "the ideal of progress" and anticipates communal action. However, in Silko's novel, the community of action hopes to precipitate the destruction of the existing world and designs alternative futures for a postapocalyptic order.

Almanac thus corroborates the persistence of apocalyptic rhetoric into the late twentieth century, and invites Lawrence Buell's question: why is "apocalypse . . . the single most powerful master metaphor that the contemporary environmental imagination has at its disposal"?[82] One obvious reason in Silko's case is that both Native and non-Native U.S. literary history is replete with millennial models—from the Mayan and Yaqui sacred works that are explicit intertexts for *Almanac* to the jeremiad mode that threads through American letters. Invoking the jeremiad, a literary tradition with a Protestant Christian heritage, does not discount *Almanac*'s debt to tribal belief. Because Indian cultures are not the only ones with which Silko is in dialogue, we must account for the novel's many positions within U.S. literary history. The three-hundred-year-old literary jeremiad offers a vision of regeneration out of degeneration and is another significant template for the novel's apocalypticism. Like Killingsworth and Palmer's millennial mode of environmental discourse, the jeremiad motivates communal action via social critique. In his influential 1978 study, Sacvan Bercovitch argues that the rhetorical mode constructs a narrative bridge between decline and renewal for its colonial American audience; it is "designed to join social criticism to spiritual renewal, public to private identity, the shifting 'signs of the times' to certain traditional metaphors, themes and symbols."[83] Whereas Perry Miller classifies it as a pessimistic genre that registers the Puritans' "'descent into corruption'" and social and spiritual decay, Bercovitch contends that the jeremiad, despite its apocalyptic patina, is fundamentally optimistic.[84] It traffics in the tropes of Armageddon to drive a triumphant narrative: out of a debased present, the community can fulfill the promises of its ideal image as found in Scripture. The jeremiad ratifies and institutes a "divine plan of progress" for citizens of the City on the Hill and confirms that God's providence and the features of incipient U.S. capitalism are compatible.[85]

With its message denouncing American profligacy and its means of delivering this message, *Almanac* adapts the jeremiad, as Native writers such as the Pequot preacher and politician William Apess had in the nineteenth century.[86] The fragments of the Mayan almanac circulating in the novel foretell the restoration of an ideal age even as they condemn present iniquity, and the speeches that Clinton, the Mayan revolutionary Angelita, and others deliver reinforce this message through fiery oratory.[87] While denouncing the destruction European colonizers initiated, *Almanac*'s revolutionaries share in a "vision of the future" around which a community coheres,[88] and impassioned speakers disseminate this message in person or via late twentieth-century broadcast technologies. The novel's prophetic structure and the exhortatory rhetoric that provokes anxiety make the novel a delayed cousin of the jeremiad. However, the novel's censure of capitalism and colonialism also align it with the "anti-jeremiad," a mode of nineteenth-century U.S. writing that arose as authors began "reading into America the futility and fraud of hope itself."[89] Although Silko populates her text with arrivistes who have ideological and economic commitments to ideals of scientific progress and class mobility—we can count Venezuelan architect Alegría, Trigg, Leah Blue, and Menardo in this group—their projects are precisely the instruments *of* corruption and doom and not tools to be used *against* them. Adapting the jeremiad as a rhetorical strategy, Silko's text denounces the Euro-American capitalist variant of progress as "fraudulent" and advances an alternative version of hope: the fulfillment of crisis, revolutionary conflict, and the restoration of tribal lands. To Buell's question, "Can our imaginations of apocalypse actually forestall it, as our fears of nuclear holocaust so far have?" *Almanac* suggests that they cannot.[90] While in some cases apocalypticism is prophylactic and meant to spur social reform, here it is predictive of planetary upheaval.[91] The narrative refuses a homeopathic therapy for the pervasive crisis it depicts. In other words, what brought the culture to the current precipice—capitalist and neocolonial expansion and technological exploitation of human bodies and the earth—will not draw us back from doom.

Like the jeremiad mode that it advances, *Almanac* makes "anxiety its end as well as its means."[92] Anxiety is the tone of apocalyptic rhetoric, its "affective bearing, orientation, or 'set toward' the world."[93] This chapter has thus far detailed the biotechnological plots and medicalizing metaphors that combine to produce this affective tone. It remains to consider what attributes of anxiety might make it an appropriate helpmate to Silko's particular apocalypticism. With traction on this question, we can address whether apocalyptic thought and anxiety, its affective tag, are compatible with social action in general and revolution in particular.

The affects that I analyzed in foregoing chapters—discord, wonder, and disgust—are shaped by antecedent normative, cognitive, and epistemological experiences, but are grounded in the present moment of encounter. By comparison, anxiety bends toward the future. It arises from a totalizing image of the now that anticipates further degeneration of the present in a foreseeable future. This idea endorses but also revises Ernst Bloch's definition of anxiety as an "expectant emotion" that speaks to "the Not-Yet, of what has objectively not yet been there."[94] Forged in the past, the present cannot imagine an alternative and so looks out into the future through anxiety. The affect is one "whose drive-intention is long-term, whose drive-object does not yet lie ready, not just in respective individual attainability, but also in the already available world."[95] Because the outcome of anxiety is not ready to hand, the affect "open[s] out entirely into this horizon" of the to-come.[96] In response to Bloch, I emphasize that the future cannot be anxiety's only temporal orientation because present conditions of the "already available world" guide the negative anticipation of the unknown. It is not just the imminence of the future that ramps up anxiety, converts it into a "'more burning'" passion;⁹⁷ it is also the immanence of radical change that causes anxiety to inflect the present. Though Sianne Ngai acknowledges anxiety's temporal dimensions, she is more concerned with the affect's spatial ontology. She bases anxiety's aspect of anticipation in a Heideggerian state of being "thrown, hurled, or forcibly displaced."⁹⁸ One's "initial surrender to being 'thrown,'" one's ultimate passivity, provides an opening to utter "absorption in sites of asignificance or negativity."⁹⁹ The spatial form of anxiety is forced distance that, despite an unsolicited projection, puts space between the subject (a "knowledge-seeking" male, in the cases Ngai examines) and threats to him.¹⁰⁰

Bloch's and Ngai's accounts provide the three crucial terms—one temporal, the other two spatial—for analyzing anxiety's effect on action in Silko's ecosickness novel. *Almanac*'s anxious tone arises from its doubled orientation toward a debased present and even worse future as well as from its indefiniteness and positions of passivity. All of which have important consequences for its ability to activate revolution. A correlate to the text's apocalyptic form, anxiety manages response to the present structure of relations by projecting their resolution—in this case, for both better and worse; indeed, for better after worse—into a disrupted future. The pending catastrophe is one of the "sites of asignificance" to which Ngai refers because apocalypse cannot be totally known and is a state of total negativity, that is, the destruction of all that exists. In arguing thus, I do not endorse a view common to Kierkegaard, Heidegger, and Bloch that anxiety is oriented to nothing.¹⁰¹ Bloch explains that anxiety's object is amorphous. "Many only feel confused. The ground shakes, they do not know why and with what. Theirs is a state of anxiety; if it becomes more definite, then it is fear."¹⁰² (So we can say that the prospect of neuro-implantation surgery produces fear in Connie in *Woman on the Edge of Time*, while her reaction to the "dark journey" of a life of oppression produces anxiety [23]). *Almanac* helps us rethink the indefiniteness that yields anxiety in terms of the pervasiveness and magnitude of a threat rather than of its mere existence. The anxiety of apocalypse and its motivating conditions must match; they both must be complete. To manufacture totalizing menace, *Almanac* multiplies rather than retracts potential sources of anxiety.

Biotechnological interventions that alter life itself are epistemological and ontological dangers because they can reconfigure human bodies and the more-than-human world and the sociocultural relations on which these domains depend. Biotech threatens a future of nothingness: the elimination of races, species, and a habitable planet. Yet it is not in itself nothing. Rather, invasive technologies build on "the ideal of progress" but also pervert that ideal and thus produce anxiety about human ingenuity itself.[103] If we understand anxiety as having *too much* as its object, rather than "no concrete thing" as in some theorists' accounts,[104] we can answer for *Almanac*'s hyperbole and other strategies of intensification and multiplication.

The effects of this too-muchness are significant and are visible in readers' responses to the novel. Moral, psychic, and corporeal registers intermingle in reviews that attest to the novel's affective impact. One reviewer's dismissive comment that *Almanac* amounts to "clumsy comic book fare" is an exception; most others detail the unignorable, embodied reactions that the text produces.[105] *Almanac* is a "vivid, preposterous, splinter-under-the-fingernails book [that] is guaranteed to make you mad and just as sure to make you squirm," according to Malcolm Jones of *Newsweek*.[106] For Brad Knickerbocker, it is "so dark, indeed perverted—to the point where some passages are exceedingly painful to read."[107] The book itself announces its corporeal impact: the prefatory pages of the 1992 edition include William Kittredge's compliment that the novel is "electrifying [and] tough to swallow."[108] *Almanac* is stirring. Charged with the negative affects of a "dark," "perverted," nigh "preposterous" contemporary American society, it charges readers up and moves them to states of madness, irritation, and pain. Indeed, it even compels them to move ("squirm").[109] But this is a contained movement. Squirming—moving but in place—is a verb that fittingly describes how an affect that arises from expansiveness can in fact strap the reader to his chair and immobilize him, even as it might prod him to anxiety over crisis.[110]

"If an entire culture reeks of cruelty and death, if the individual imagination confronts not a specific loss but fathomless brutality, if despair is no longer a matter of private experience but pervades an epoch," the situation in *Almanac* as Elizabeth Tallent describes it, action is impossible.[111] Despite provoking visceral reactions, the novel risks etherizing the reader. *Almanac*'s momentous call for the total overthrow of white rule in the Americas and return of the land to the dispossessed works through hyperbole and the anxiety of apocalypticism, but the very largeness of these narrative techniques can just as easily obstruct kinetic response as galvanize it. In this respect the novel contends with a strategic problem that confronts environmental activists as well. Frederick Buell comes to this point in his analysis of crisis discourse in environmentalism. "A total crisis of society" he worries, "helps keep merely dysfunctional authority in place.... It depoliticizes people, inducing them to accept their impotence as individuals; this is something that has made many people today feel, ironically and/or passively, that since it makes no difference at all what any individual does on his or her own, one might as well go along with it."[112] One contingent of reviewers dismissed *Almanac* for discouraging readers through its excess, but rejecting the novel is shortsighted. It makes a significant contribution to environmental thought by revealing the limits of the various forms that environmentalism takes—from affirmation of human-nonhuman fusion to racialized populationism and doomsday alarmism. Moreover, *Almanac* shows how an affect like anxiety contributes to those limits and arises from particular narrative strategies. Whereas the Mattapoisettans' agrarianism in *Woman on the Edge of Time* is utopian, Silko's novel asks us to contemplate an even more radical break from present reality. In order to relay the pervasiveness of present injustice, it imagines where we're ultimately headed: toward the reconfiguration of life as we know it and, thanks to biotechnology, of life itself. The text urges that reconfiguration is inevitable in order to establish the foundations for the revolution that it foretells. In doing so, it foregoes preemption in favor of anticipation. What Ngai reads as anxiety's "initial" position of "impotence or passivity," however, is the affect's final position in *Almanac*.[113] Anxiety suspends text and reader between the prophylactic and the inexorable, between the knowledge that action is necessary and a feeling that we are powerless to redress systemic social, bodily, and environmental sickness. Even though Piercy's novel ends with radical uncertainty about whether the utopian future will come, the coexistence of the pre- and postrevoluntary worlds within the diegesis offers the reader a choice rather than suspending her in the present and risking passivity.

My claims about the suspensions of anxiety introduce a series of questions. Is passivity a by-product of apocalypticism and its affective signature, anxiety? Or does it thread through the narrative by other means? Addressing these queries directs us to the forms of agency that *Almanac* assumes and its characterological solution to the suspensions of anxiety. Ultimately, I argue that not only does plot generate affect; affect also puts pressure on plot.

SQUIRMING AND TREMBLING

"'You think there is no hope for indigenous tribal people here to prevail against the violence and greed of the destroyers? But you forget the inestimable power of the earth and all the forces of the universe. You forget the colliding meteors. You forget the earth's outrage and the trembling that will not stop'" (723–24). Individuals might only squirm, but, according to Wilson Weasel Tail, the earth itself will tremble when "the sun burns the earth" (724). The disproportion between one critic's physical reaction to *Almanac of the Dead*'s apocalypticism—"squirming"—and the geophysical forces that herald the apocalypse—"trembling"—reveals a discrepancy between two forms of agency circulating in the novel, one relating to technoscientific change and the other to historical change.

For biotechnologists, there are no limits to human choice and action. For the novel, this means that there are no limits to the injuries that interventions into life can inflict. As day broke on genetic technologies, the horizon of choice elongated. With greater manipulation of the human genome came ebullience about the opportunity for self-determination. Nikolas Rose builds his "politics of life itself" around this biotechnological extension of individual choice. He recognizes that genetic technologies potentially carry the taint of eugenics, but he ultimately celebrates the "new vital politics" and techniques for optimizing one's life.[114] He reassures readers that "we are not on the verge of a new eugenics, or even a revived biological determinism. . . . In the new field of biopolitics, where interventions are scaled at the molecular level, biology is not destiny but opportunity."[115] To shepherd these technologies into everyday life "is not to resign oneself to fate but to open oneself to hope."[116] In this respect, Rose's position on biotechnology precisely inverts *Almanac*'s anxious stance. This is because hope shares anxiety's expectant form but reverses its content, declaring "Yes to the better life that hovers ahead."[117] With its anxious affect, *Almanac* sees biotechnological agency as serving the wealthy and powerful at the expense of the marginalized. To put it more strongly, hope for the oppressed diminishes in the novel in inverse proportion to the magnitude and scale of individual choice. The thrill of opportunity excites technological optimists, but *Almanac* converts excitement into apprehension. It exhibits anxiety at the very inventions that embody (some) individuals' will.

Yet *Almanac*'s stance toward human agency wavers depending on whether humans are imagined as biotechnological innovators or as instruments of historical change. Whereas technologists like Serlo and Trigg have the power to manipulate and manage the stuff of life through their megalomaniacal schemes, geophysical forces trump individual actions when it comes to revolution. In this way, the contradiction of the apocalyptic cultural form plays out in *Almanac*. On the one hand, the novel suggests that "it makes no difference at all what any individual does on his or her own" while, on the other hand, it "connect[s] an individual vision to a shared future."[118] In the technoscientific realm "individual vision" is itself apocalyptic; it is a way for elites to exacerbate inequalities, perpetrate violence, and threaten human and nonhuman existence. Serlo's, Trigg's, and Leah's projects instance this point, and the private arrangements that bring them about—a remote ranch, commissioned research, backroom dealings—highlight the individualism of their endeavors. Undoubtedly, the novel also envisions how people such as artists and thinkers might steer the positive remaking of the world, as Lois Perkinson Zamora's study points out. However, *Almanac*'s revolutionary historiography is inconsistent and ultimately cinches human potential to effect change. It both makes history synonymous with charismatic agents and announces that the earth itself, not those who people it, brings about revolution.

Other readers of *Almanac* have noted the text's conflicted portrayal of revolution. T. V. Reed remarks that "Silko's evocation of the coming revolution offers no blueprints or predictions as to how events will unfold. In the novel and in interviews, she offers multiple, contradictory, equally possible versions of the form the coming revolution will take."[119] Silko doesn't see this inconsistency. In one of those interviews to which Reed refers, the author confidently asserts that *Almanac* has a univocal message about how the sicknesses that it narrates will be cured. Commenting on her vision of revolution, Silko explains, "As I tried to make clear in *Almanac of the Dead*, you don't have to do anything, for the great change is already happening. But you maybe might want to be aware of what was coming, and you might want to think about the future choices that you might have to make. Though as I said, in your heart, you will already know."[120] Silko's slide between tenses underscores the multiple time stamps on the novel's predictions. For some characters, the revolution is at hand, for others it awaits, and still others see it as immanent in all acts of rebellion throughout history. The restoration of tribal sovereignty and the healing of people and land are never in doubt in the text. Nor in life, as Silko explains in the 1994 interview: "all those criminals in the United States Congress—their time is running out very soon! The forces from the south have spiritual power and legitimacy that'll blast those thieves and murderers right out of Washington, D.C."[121] But the sources of the actions that will effect change are ambiguous. Ultimately, Silko and her novel restrain people's role in bringing about revolution and assign it to more-than-human agents. She continues:

> Ah, ah, this change that's coming will not have leaders. People will wake up and know in their hearts that it's beginning.... We don't need leaders. They can't stop [this revolution]. They can shoot some, they can kill some—like they have already—but this is a change that rises out of the earth's very being—a Hurricane Andrew, a Hurricane Hugo, an earthquake of consciousness. This earth itself is rebelling against what's been done to it in the name of greed and capitalism. No, there are no groups which bring change. They aren't needed.[122]

To place all hopes for change in a single leader or group threatens change itself. This is evident as the term *leaders* shifts from referring to the shepherds of change to those who want to halt it. Having evacuated the human element, Silko promises that "this earth itself," its geophysical forces of destruction, will be its own savior.[123]

Silko's characters give voice to similar accounts of a revolution the planet itself initiates to root out sickness and injustice. Calabazas, a wizened Mexican-Yaqui drug smuggler, iterates various tribal prophecies predicting that cataclysmic meteorological events will effect the overthrow of white power: "In each version [of the prophecies], one fact was clear: the world that the whites brought with them would not last. It would be swept away in a giant gust of wind. All they had to do was to wait. It would be only a matter of time" (235). The Barefoot Hopi adds "earthquakes and volcanic eruptions of enormous magnitude... tidal waves and landslides. Drought and wildfire" to the litany of forces that herald earth's vengeance (734–35). It is not only Native characters who share the geophysical theory of revolution. Clinton echoes Calabazas's prediction that a sea change is afoot: "Great winds would flatten houses, and floods driven by great winds would drown thousands. All of man's computers and 'high technology' could do nothing in the face of the earth's power. All at once people who were waiting and watching would realize the presence of all the spirits" (425). Silko's accounts of revolution in her novel and in interviews show a certain antihumanism that is also apparent in the nearly universal degeneracy of *Almanac*'s characters. While people, especially whites, can destroy the planet and their fellow humans under the sway of greed, hatred, and lust, the earth and its spirits are the curative agents against sickness and injustice. If the earth's forces or numinous spirits will pilot the revolution, why not just be a passenger, sit back, and "go about... daily routines" (735)?

In addition to promoting passivity, the geophysical theory of revolution is in tension with *Almanac*'s idea that strong personalities drive history. Just as individuals propel the current trajectory of biotechnological development, they also shape the actual past (as opposed to the anticipated future). Through the novel's sentence grammar and in its scenes of history telling, totemic power accrues to proper names.

Open up *Almanac* most anywhere and individual characters' names leap off the page because of their incantatory repetition. To take just one example, "Trigg" appears eighteen times on page 379 of the 1992 Penguin edition, and in eleven of those cases it is as the subject of the sentence. This stylistic habit gives this paragraph and many others like it an anaphoric effect.[124] Grammatically, then, the individual is a potent subject and agent in the narrative. Furthermore, in *Almanac*'s historical imagination people are synecdoches for pivotal events. The Army of Justice and Redistribution executes the Marxist Cuban Bartolomeo for "crimes against history" that are primarily crimes of nominative omission (313). La Escapía (Angelita's nom de guerre) and El Feo order his killing "for other crimes too," but "they remembered that mainly Bartolomeo the Cuban had lost his life because he had neglected to mention the great Cuban Indian rebel leader Hateuy" (315). Omitting the leader's name from a historical narrative is tantamount to erasing history itself. While "stories of depravity and cruelty were the driving force of the revolution," individuals are the driving forces behind revolutionary histories (316). Grammatically, historiographically, and, finally, diegetically, the novel is remarkably peopled, and peopled above all with would-be leaders and catalysts of change. Angelita, Bartolomeo, El Feo, the Barefoot Hopi, Zeta, Trigg, Leah and Max Blue, Serlo, and Beaufry: these are all characters whose projects for altering present and future realities the novel details. As it excavates the personal motivations and the idiosyncratic beliefs that steer these schemes, the text asserts that individuals determine the future and are not merely stand-ins for rapacious capitalism, in the one case, or grassroots rebellion, in the other.

My analysis of the contradictions of agency in *Almanac*'s revolutionary theory substantiates but modifies Walter Benn Michaels's argument that Silko's text "replace[s] the differences between what people think (ideology) and the differences between what people own (class) with the differences between what people are (identity)."[125] Michaels adduces *Almanac* in his argument that recent U.S. novels and literary theory have transformed "subjectivity into subject position," that identity politics constitute reader experience as the final basis for textual meaning.[126] "The only thing that matters" in the readings he lambastes is "the subject position of the reader," in particular the collective memory that is inscribed on her mind and body, and the kind of experience that that position authorizes.[127] When the Army of Justice and Redistribution executes Bartolomeo, Michaels holds, they transmute identity into history as communal memory.[128] To blend my reading with his, nominative omission merits death because the individual, in this case Hateuy, stands in for a history that confers identity onto a community imagined not as a class but as repositories for a "historical legacy."[129] To both of our minds, *Almanac* evacuates the kind of individual action on which its politics would seem to depend. However, he misses two crucial points that emerge in my interpretation of sickness. His exclusive focus on whether the novel adequately critiques late capitalism and envisions the redistribution of wealth causes him to overlook the novel's inconsistencies. First, by making planetary forces the instigators of social, political, and ecological changes, *Almanac* posits revolution as independent of *both* identity as history *and* history as class struggle. Second, the novel's biotechnological imagination unmoors individuals from history even as it makes them agents of an elite-driven class struggle that deepens existing socioeconomic and racial inequalities rather than subverting them.

These inconsistencies result from a tension between the two genres announced by the book's title: the almanac and the novel. Not only does *Almanac*'s apocalyptic form suspend it between prophylaxis and inevitability; its titular forms wrench it by one turn toward the planetary and by another turn toward the individual. Both genres can safely house the apocalyptic mode. However, the almanac, a genre of "astronomical data and calculations, ecclesiastical and other anniversaries, besides other useful information, and, in former days, astrological and astrometeorological forecasts," exerts a gravitational pull toward planetary explanation and prediction.[130] In the text these predictions point to the apocalyptic. As Lecha clarifies, "'Those old almanacs [in this case, the Mayan and Yaqui texts that she is preserving] don't just tell you when to plant or harvest, they tell you about the days yet to come—drought or flood, plague, civil war or invasion'" (137). Yet the book also stands behind its status as a novel, even declaring it in its subtitle. *Almanac* therefore invites the expectation that it will detail characters' lives, thoughts, and conflicts that here occur at interpersonal, tribal, national, and transnational scales. *Almanac* is ultimately torn between these genres and between multiple accounts of human and planetary agency.[131]

The novel attempts to seal up the fissures that its generic leanings and anxious affect open up, but this only further bends the trajectory of the ecosickness narrative. It tries to change the emotional character of the novel by changing the use of character. This becomes clear in the concluding chapter, "Home," which is poised between a present of endemic environmental, somatic, and social sickness, and a future in which tribal sovereignty, justice, and health are restored. The chapter restates history's indifference to human action: "The [Laguna stone] snake didn't care if people were believers or not; the work of the spirits and prophecies went on regardless. . . . Burned and radioactive, with all humans dead, the earth would still be sacred" (762). The snake's thoughts seem to travel to Sterling via the wind moving through the junipers, and they carry with them echoes of Silko's extratextual comments that individuals are dispensable in the revolutionary fight for justice. The snake's confidence in the unconditional consecration of the land naturalizes the earth's triumph. Yet, despite human insignificance, the south-facing stone snake still portends the rebel's arrival from Mexico, "the direction from which the twin brothers [El Feo and Tacho] and the people would come" (762–63). These words—"and the people would come"—are the last ones in *Almanac*, and they end the novel in the conditional tense. Readers do not witness systemic healing as they do in *Woman on the Edge of Time* but are invited to fill in the "if . . ." clause that would make the hypothetical actual. The conclusion is anticipatory with respect to the revolution, but it resolves Sterling's rootlessness and restores his connection to tribal land and stories by announcing "that Sterling found it was easy to forget that world in the distance; that world no longer was true"; it "had only been a bad dream" (757, 762). The novel concludes with a single character's escapist fantasy that erases the previous 750 pages. Therefore, though the conditional tense of the concluding clause suspends the story, the narrative works to recover from this openness by focusing the plot on a characterological center. This is a departure for a text that has been centrifugal until this point. Leading up to "Home," the narrative multiplies characters and diffuses them, just as it dissipates action through anxiety. The last chapter hopes to recover from the openness of biotech anxiety, the suspension of agency, and the dispersion of character by abandoning *Almanac*'s large cast and foregrounding a healed subject.¹³² It does in its conclusion what *Woman on the Edge of Time* does throughout: produce a more intimate narrative as an antidote to pervasive threat.

With a feeling subject at the narrative center, the negative affect of anxiety evaporates. In Ngai's reading of anxiety through Alfred Hitchcock's *Vertigo*, Herman Melville's *Pierre*, and Martin Heidegger's *Being and Time*, the emotion "rescues," "restores," and "validates" the questing protagonists—indeed, it can produce "a form of 'revolutionary uplift'"—to the extent that there *is* a characterological focus.¹³³ At its end, *Almanac* seems to learn that this is the escape route from what Ngai terms anxiety's "directionless oscillation" and therefore instates an individual as prospective witness to global environmental and sociopolitical renewal (and erasure) that mimics his personal restoration.¹³⁴ *Almanac*'s conclusion shows in miniature the pull between enthusiasm for the individual actor and anxiety about what powerful agents bring into being through biotechnological invention. The novel's account of the geophysical agents of revolution fulfills one wish: if humans are not the arbiters of the future, there is hope that the earth itself will bring about regeneration. "Home" expresses another wish: that the text's own story will melt into air, leaving only a conventional protagonist as the sentimental locus of redemption.

Both *Woman on the Edge of Time* and *Almanac of the Dead* respond to the question whether there is a place for biotechnological interventions into life itself in a socially and environmentally just future. They challenge readers to reconsider dominant narratives of technological progress by proposing that apocalypse is in fact the telos of innovation. However, the texts diverge in how they answer this question and deploy apocalypse. Piercy's novel includes apocalypse within its temporal frame, but then integrates biotechnologies into the very pattern of a postapocalyptic utopia. The anxiety of intervention in *Woman on the Edge of Time* does not inhere in life technologies, but only in the hubris and racism of their manipulators. Just as genetic engineering, for example, can change the course of plant evolution, so can democratic oversight change the course of technological evolution. Piercy mobilizes anxiety, but the novel's dual temporal structure offers relief for readers in order to make space for entertaining affirmative communalism in an immanent future. Silko's ecosickness narrative is more ambitious in scope and forces us to adjust to a world of pervasive sickness in which body and land are vectors for anxious feeling. *Almanac* posits that, because the very natures of vulnerable human bodies and environments are resources for projects that perpetuate iniquity, the manipulation of life itself cannot be redemptive.

In my reading, Silko's prescience hinges on her book's horrified vision of how technoscientific interventions can penetrate all domains of existence. Portraying the biotech colonization of life as a fait accompli, *Almanac*'s prophecy is at its most felicitous. Twenty-first-century readers know that technoscience proceeds apace in the next millennium, and the fact of biotechnologization has not burned up in the flames of a hemispheric revolution along with dispossession, greed, and inequality. Through a story that reworks environmentalist tropes of merging, conventions of apocalypticism, and the vectors of anxiety, *Almanac* at once challenges ecocentric and techno-optimistic rhetorics within environmentalism. To acknowledge that Silko's prescience rests on her biotechnological imagination as much as on her vision of territorial restoration and coalitional environmental justice activism solidifies *Almanac of the Dead*'s importance to contemporary literary and ecocritical studies. In fact, forcing the novel into the disease-to-recovery narrative arc, one set by *Ceremony*, occludes its unique value to current ecocriticism. When we instead read *Almanac* as an ecosickness fiction in which healing and restoration are always attenuated, we identify the affective range of environmental thought and writing—from redemptive body-land fusion to anesthetizing anxiety. We see how those feelings coexist between the covers of one book and strain each other. With a critical eye to the narrative affects that imbricate body and environment under the sign of sickness, we become aware of how emotion puts pressure on the activist leanings of fiction.

Conclusion

HOW DOES IT FEEL?

My days start like those of many news-hungry "internauts," with a peek at the headlines that Google Reader aggregates. There's no front page here. Google collects, and I select. Do I first unlock the folders that hold stories of international climate aid, environmental policy analysis, and the latest medical breakthroughs and warnings? Or do I scan the items under "culture": book reviews and miscellany from *TheMillions.com*, essays from *Los Angeles Review of Books*, and interviews from *The Believer*? The banality of this ordinary habit is deceptive. Seemingly about gaining quick information about the day's events, it in fact unleashes a rush of emotion that can linger for hours. Stories concerning the death or vitality of humans and the planet infuse the routine with the astonishing, and the news aggregator augments this experience as it puts before our eyes so much more than the analogue front page can hold. Let's consider one day, 10 December 2012, and one folder, "environment." What feelings will this word provoke? My eyes peruse the fragments of headlines that the Reader has so neatly organized:

"Climate change talks deadlocked"
"Doha climate talks stall"
"As global climate talks founder . . ."
"NOAA's 2012 Arctic 'Report Card' . . . Danger Signs Ahead"

The 2012 UN climate summit in Doha, Qatar has come to a close at the same time as the National Oceanic and Atmospheric Association (NOAA) has announced that the summer of 2012 was "'astonishing,'" "the warmest in 170 years of recordkeeping," with rapid ice melt that (yet again) portends wide-ranging, unpredictable climatic shifts around the planet.[1] So how does "environment" feel on this morning? At an impasse, impossible, maybe even hopeless. Undoubtedly imperiled.

The headlines on this day elicit reactions that students frequently express midway through a semester of reading recent environmental fiction. Where's the hope? In their essays on environmental pedagogy, two ecocritics who have appeared in this book, Joni Adamson and Stacy Alaimo, confirm my sense that this is an unavoidable question in the environmental studies classroom.[2] Even after reading works that extol the restorative, ennobling aspects of environmental experience— say, Leslie Marmon Silko's *Ceremony* or Barbara Kingsolver's *Prodigal Summer* (2000)—students seek hope with an amalgam of urgency and exasperation. Books by Margaret Atwood, Elizabeth Kolbert, and Indra Sinha that an ecoliterature syllabus might feature all do the good work of demystifying the naturalness of nature, challenging the paradigm of technological progress, relaying information about climate change, and making visible the disproportionate burden of ecological disaster shouldered by the poor. Some, Sinha's *Animal's People* (2007) comes to mind, even incorporate into their plots glimmers of hope in the form of grassroots activism. Yet despair still tends to snuff out that faint light for most students. And this is not because they belong to a lost generation of cynical Millennials, but *despite* their aversion to pessimism. Just as the negative review stings more than the positive one soothes, plots of injury and injustice linger longer than those of reparation and remediation. In light of this, Adamson and Alaimo highlight activist successes in their classes, an approach that environmentalist Paul Hawken takes in response to the mood of despondency in much environmental discourse. After depressing audiences at his first public lectures, he began looking to the distributed network of environmental and social justice groups combating globalized "free market fundamentalism" and its ravages.[3] Following these groups' journeys toward justice, Hawken inadvertently tarries with the positive. "I didn't intend it," he explains; "optimism found me."[4] His phrasing hints at defensiveness, as if he's initially uncomfortable with the desire for hope itself. The concern might be that optimism betokens naïveté. To intercept this judgment, Hawken presents hope as surprising and unsolicited. It arises in the manner of those affects that I have explored in *Ecosickness in Contemporary U.S. Fiction*: as unbidden as the wonder that overtakes Karin Schluter while doing research for the Buffalo County Crane Refuge in *The Echo Maker*, perhaps even as shocking as the discord that revises Jan Zita Grover's and David Wojnarowicz's conceptions of nature, health, and beauty.

At the end of the long work of researching and writing a book on contemporary environmental fiction, I wonder: does optimism have a place in this project? More importantly, does the reader find optimism, or does this book send its audiences on the same emotional ride as the typical environmental studies course? Does the uneasy question "whither hope" surface without an easy answer? Before I delve into these questions, it's worth returning to Hawken's discomfort with hope. The conversation around hope within environmental discourse is itself anguished, as some voices endorse hope as a resource for activism and others as a hindrance to it. As I have mentioned, one reason for skepticism toward hope is that it can mark the one who believes in a better future as naive and ignorant in an age when "hip cynical transcendence of sentiment" is still endemic to certain demographics.[5] Another reason is that within environmentalism the most vocal form of optimism is techno-optimism. For many environmental thinkers—the authors in this study included—a futurity based primarily on technological innovation amounts to Western capitalist business-as-usual. Smart grids, smartphones, and Smart cars alone won't deliver us from our "dumb" ways of living so much as perpetuate them.

Optimism thus makes environmental advocates wary even as it drives them, because it opens one to accusations of ignorance or delusion and because it is associated with technological entrepreneurialism. We can add to these reasons more complex psychological and historical ones that reintroduce the concern that has motivated the preceding chapters: how affect bears on the ethics of ecosickness. Might there be dangers to optimism? And, to backtrack further, might we actually have a surfeit rather than a deficit of it?

Two arguments point to the perils of optimism: what brain scientists term *optimism bias* and what journalist Barbara Ehrenreich identifies as an American penchant for positive thinking. In a November 2011 study published in *Nature Neuroscience*, a team of researchers investigate the neurobiological bases for optimism bias. This is the tendency for individuals to "updat[e] their beliefs more in response to information that was better than expected than to information that was worse."[6] In effect, we hold to beliefs that herald a promising future for ourselves despite evidence suggesting that we are in fact vulnerable to disease, crime, and other harms. Citing the paper, a writer at *Grist.org* speculates that this bias is one of many reasons "why we can't do anything about the existential crisis of climate change—or, indeed, any of the other existential crises we're facing at present."[7] The biologists seek a neurophysiological basis for a phenomenon Ehrenreich locates in American history and social formations. The spark for her investigation is the "pink sticky" positivity into which women are recruited once they receive a breast cancer diagnosis.[8] The roots for the Susan G. Komen Foundation's pink ribbon enterprise lie in the early to mid-nineteenth century, Ehrenreich contends. This is when positive thinking arises as a reaction to patriarchal "Calvinist gloom."[9] As the gains of U.S. imperial expansion were trumpeted, and thinkers such as Ralph Waldo Emerson queried "'why should we grope among the dry bones of the past. . . . Let us demand our own works and laws and worship,'" a new study of happiness took shape and rethought the Calvinists' harsh God.[10] Both national and personal promise shone brighter, and American institutions and people embraced the upbeat.

And what's wrong with bright-sidedness, we should rightly ask? The *Nature Neuroscience* paper and Ehrenreich's story of the birth and dominance of U.S. optimism pinpoint the downsides of looking up. First, optimism prevents us from recognizing signs of adversity for which we could prepare in preference for a narrative of triumphalism that in fact precipitates danger. In this view, captured by Richard Powers in a 2011 essay, "it won't be our capacity for despair that does the race in; we are damned by how easily we shrug the darkness off."[11] Second, optimism walks hand in hand with the capitalist mandate to grow at all costs and, in Ehrenreich's estimation, obfuscates "the crueler aspects of the market economy" in favor of "a harsh insistence on personal responsibility."[12]

The dangers of positivity readily pertain to environmental threats such as species extinction, water depletion, and climate change. Thinking positively, we assume that individual consumerist behaviors, market and technological innovations, and/or simple good fortune will keep loss of life and property at bay. Just as environmental citizens and students yearn for hope, then, hope can be a stymying affect. Even so, is it not better to cultivate the redemptive narrative that Marge Piercy sows in the Mattapoisett plot of *Woman on the Edge of Time* than to remain in a fallow state of nihilism? Without optimism, aren't we forever inert?

This is a question that queer affect theorists such as Ann Cvetkovich and José Muñoz puzzle over. In *Depression* Cvetkovich inventories the embodied personal, scholarly, and political practices that "help me get up in the morning" when "saying that capitalism . . . is the problem" can make even the most stalwart activist want to crawl under the covers in a state of political despondency.[13] Rather than drawing on reserves of positive thinking in the manner of the pink ribbon brigades, Cvetkovich uses negative feelings like depression "as a possible resource for political action rather than its antithesis."[14] For her, as for Muñoz, positive thinking passes by way of the negative and can thus bypass bias, exceptionalism, and delusion. Their optimism involves using negative affects "in the service of enacting a mode of critical possibility."[15] They promote a third way to crippling cynicism and stultifying optimism.

This range of thought on optimism—neurally based bias and historically contingent exceptionalism, utopianism of the ordinary and positivity by way of the negative—underscores one of the insights of *Ecosickness*: just how messy ethicopolitical emotions always are. The most basic reason I set out to write this book was to account for the tropes of sickness that recur in environmental writing in our technoscientific present. But I soon found that in order to theorize sickness I also had to account for the astonishing variety and, more importantly, unpredictability of emotions attached to environmental and somatic experience in the late twentieth and early twenty-first centuries. The writers that I have gathered here turn to affect under urgent conditions: the AIDS epidemic for Grover and Wojnarowicz, brain injury and habitat and species loss for Powers, endemic detachment and toxification for David Foster Wallace, and social injustice and revolution for Piercy and Silko. The environmental dispositions that arise under these conditions undoubtedly depend on preexisting social norms, as I discuss in chapters 1 and 2, but they also take shape around the jolt of everyday encounters with human and nonhuman others: the unexpectedly beautiful clear-cut (*North Enough*), the grotesque toxic body (*Infinite Jest*), and even the miserable data of habitat loss (*The Echo Maker*). Essentially, ecosickness narratives attest that crises of bodily and planetary endangerment are also affective crises. Rather than conclude that sickness indexes collective despair at a lost cause, however, I pondered how, in affiliating soma and earth, environmental thinkers use sickness as a way to involve readers in the fates of both. The writers that I have examined construct their fictions around the idea that, while displaying data to establish correlations between environmental damage and human disease is necessary to changing environmental and biomedical conditions, the real work of fiction lies in reconfiguring perceptions of these domains through new metaphors, tropes, stories, and emotions.

Ecosickness argues that contemporary fiction alters relationships to lived environments and conceptions of agency by drawing together ecological and somatic sickness through narrative affect. Sickness establishes structural, aesthetic, and affective homologies between changeable bodies and changeable ecologies and thus breaks down the human-environment boundary. These homologies arise in ecosickness novels and memoirs through plot and narrative perspective, play with literary conventions, and description. On this last feature, the medicalization of space not only registers the increasing biomedicalization and technologization of life but counteracts it by insisting that we are not isolated bodies, "machines in space, composed of machine parts."[16] To perceive death in a mountain (*Close to the Knives*), to detail urban infrastructure through human physiology (*Infinite Jest*), and to see aging skin in a cholla cactus (*Almanac of the Dead*) are, for my authors, ways of apprehending the inseparability of our somatic and ecological fates. To return to the questions I raised previously about where this project leaves me and the reader, it's safe to say that these writers are by no means universally optimistic, but they propose that this inseparability is the key to any possible hope.

This book is then cautiously optimistic insofar as it illuminates contemporary writers' attempts to ferry readers to environmental consciousness, even if on the raft of sickness and vulnerability. Its hope rests on representational and conceptual intimacies of earth and body rather than on the idea that there is one felicitous emotion that catalyzes alternative futures. *Ecosickness* echoes Gus Speth's declaration in his manifesto for a socially, economically, and environmentally just America that "now we need to hear more from the preachers, the poets, the psychologists, and the philosophers,"[17] and it shares Powers's position on why this is the case: because "narrative imagination can twist our guts and shatter our souls."[18]

But in what directions and into how many pieces is never entirely countable. Uncertainty about the outcomes of affect makes it hazardous terrain, for the artist and for the critic. Ecosickness fiction is a literature of accountability because its narrative and affective techniques show the networks of obligation to which we belong. Yet these authors do not perforce believe that literature is a perfect conversion engine that turns images and stories into biomedical and environmental knowledge, ethics, and politics. Such a conversion will sometimes take place, without a doubt, but this is not the promise of ecosickness fiction. What it promises instead is to use images and stories to set into motion the messy emotions that can alternately direct our energies toward planetary threats and drive them away from action. In these narratives anxiety can lead to suspensions of agency even in the midst of revolution (*Almanac of the Dead*), and wonder can shade into paranoia rather than care (*The Echo Maker*). But, then again, discord can validate experiential authority (*Close to the Knives* and *North Enough*), and disgust can aid attachment (*Infinite Jest*). Ecosickness narratives might attest that our understanding of affect's effects is provisional, but they trust that it is emotion that can carry us from the micro-scale of the individual to the macro-scale of institutions, nations, and the planet.

Sickness is a powerful organizing concept for contemporary writers for all the reasons I have discussed. It is equally generative for contemporary literary criticism because it opens up a mode of analysis that lends specificity to concern with interconnectedness that animates cultural studies of the environment and of planetarity more broadly.[19] Ecosickness narratives show that it is impossible to approach somatic and ecological injury as isolated phenomena. We must take methodological inspiration from the literature we analyze and bring different ways of knowing—from scientific experiment to embodied feeling—to bear on each other. Following this procedure, interconnectedness becomes method and not only theme or aspiration. This is not tantamount to celebrating interconnectedness, however. Studying sickness also makes it clear that the interpenetration of self and other is as terrifying as it is enabling. Furthermore, the sickness rubric serves contemporary literary study by promoting affiliations that extant organizing categories can occlude. The authors I've considered here can fit neatly into the pigeonholes of feminist utopia (Piercy), indigenous cultural history (Silko), or "misery" memoir (Grover and Wojnarowicz), for example, and can invite comfortable interpretations associated with those categories. Yet, by reading outside these rubrics, we draw out shared efforts to represent, adjust to, and even reverse the alterations to life itself occurring today.

Even as ecosickness affiliates David Wojnarowicz and David Wallace, it does not suggest that a particular environmental or somatic politics is shared between them. We discovered this in reading Silko's and Piercy's novels together in chapter 5. They both depict interventions into the human body and the earth as shared catastrophes but part ways on the question whether biotechnology is inherently inimical to justice. Similarly, even though Grover and Wojnarowicz unlink the conceptual chain connecting nature, harmony, health, and beauty, *Close to the Knives* does not share *North Enough*'s ethic of embracing the as-is. These observations reignite the questions with which I concluded chapter 4: must environmental literature worthy of the name serve environmental*ist* ends? Must we walk away with at least a sparse blueprint for behavioral and policy changes or is depicting the affective messiness of ecosickness sufficient?

The answers to these questions speak to the identity of the field of ecocriticism, and they require us to reflect on ways of reading rather than on taxonomies of literary features. A text ripe for ecocritical study must certainly exhibit some self-consciousness about how it engages tropes of nature, depicts built and nonbuilt space, theorizes life itself, and addresses threats to the planet. However, by and large, the books in this study do not dictate an activist agenda so much as elaborate the challenges to action that our habits of thought and feeling present. Reading a text's environmentalism in these habits rather than in any explicit agenda helps us avoid the theory versus praxis debates that have misleadingly polarized ecocritics, a debate that also maps onto that between experimental form and realism.[20] Ecosickness fiction suggests that environmentalist praxis and lived relations to the environment more generally rest on an affective substrate. Approaching praxis thus requires a theoretical understanding of emotion, and this understanding arises through critical engagements with representational genres and devices. In this respect, then, a text's environmentalist status falls as much on the reader and critic as on the author. The works that this book has considered do not outline the kinds of programs and campaigns one finds in Speth's *America the Possible* or on the pages of the 350.org Web site. Instead, they consider how affect and the cultural templates that activate it either advance or block the transformations that these parties promote.[21]

One benefit of the analytical categories of sickness and affect is that they put literary fiction in conversation with these environmental activists as well as other cultural producers. Surveying ecosickness discourse within these arenas reinforces the importance of narrative affect to environmental and somatic awareness. As we saw in chapter 4 with WWF's slimy fishman and Plane Stupid's splattered polar bears, bold environmentalists are depicting mutable, vulnerable bodies to trigger disgust and horror at inaction on the climate crisis. They are experimenting with how the more-than-human makes us feel and are using emotions to drive their twenty-first-century environmental agendas. Two more groups spanning activism, art, and scholarship exemplify the trend to explore how affect-laden stories and tropes change environmental consciousness in the public sphere: Critical Art Ensemble (CAE) and the Cultural Cognition Project.

Working at the intersection of participatory performance art, lab research, and public education, the avant-garde collective CAE represents a provocative effort to probe the affects that arise as biotechnologists intervene in life itself. In their 2001–2003 *GenTerra* project,

CAE invites audience-participants to assess the risks and benefits of producing synthetic organisms that might have unintended effects on the ecosystems and human bodies that they touch. For the project, CAE thought up a biotech company whose mission is to invent "Transgenic Solutions for a Greener World," in particular a genetically spliced bacterium that can clean up oil spills.[22] During the performances, "technicians"—in fact, CAE artists in lab coats—taught lay participants about recombinant DNA and then drew samples of their blood to cross with bacteria strains. Finally, the technicians gave participants the option to release the created organisms into the environment. On the companion Web site, CAE ventriloquizes the sanitized rhetoric that biotech start-ups use to market their ventures. It attempts to excise negative emotion in order to pitch the multiple ways in which transgenic organisms might heal and protect the planet. The title of CAE's position paper on *GenTerra*, "Fear and Profit in the Fourth Domain," announces the centrality of affect to discourse on interventions into human and nonhuman life. The faux biotech firm must face public anxieties about the outsize consequences of the "categorical mixing" of human and bacterial DNA.[23] In a covert parodic mode, the art project and manifesto entertain a serious question for environmental and health activists: are strong emotions and measured decision making compatible? "Fear and Profit" ultimately concludes that the fog of fear clouds risk assessment. Countering what I've called the anxiety of intervention into life itself, CAE asserts that "the mythic past and the sci-fi future have to be separated from the reality of current research initiatives."[24] The group's online materials and performances thus build up a patina of neutrality—lab uniforms, scientific instruments, research data—to offset the emotions of fear, anxiety, and even horror that genetic modification elicits. Yet, in their approach to real scientific experiment through parodic performance, CAE in fact highlights the affective residues that always remain in biotechnological discourse and tinge the ethics of environmental, somatic, and technological change despite the best attempts of corporate marketing machines to wipe them clean.

The Cultural Cognition Project at Yale Law School takes one step back from CAE's point of entry into exploring risk's emotional entailments. If CAE's *GenTerra* stages situations where individuals must measure the dangers and advantages of corporate genetic modification, these social scientists investigate the preexisting values that inform individuals' positions on such scenarios where there are "disputed matters of fact."[25] The scholars associated with the project define these values as cultural cognition, the "disposition to conform one's beliefs about societal risks to one's preferences for how society should be organized."[26] Individualism and communitarianism and hierarchism and egalitarianism are the primary preference categories that they schematize. While there is much to say about this research, the main insight I take away from the report and the project at large is that individuals' dispositions toward public policies on biotechnology and environmental threats arise from their sense of how "'people like them'" would respond to a given set of facts rather than from a process of rational analysis.[27] When orienting ourselves toward policy questions such as whether we should increase the amount of nuclear power in our nation's energy portfolio, we undertake "identity-protective" measures; "avoiding dissonance and estrangement from valued groups, individuals subconsciously resist factual information that threatens their defining values."[28] The Cultural Cognition Project reinforces two points I make about ecosickness narratives that refract similar dilemmas of health, technology, and planetary endangerment: that a person's environmental and somatic knowledge comes only partially from the data of empirical research. Emotion and sociocultural values constitute a much larger part of the matrix of factors that organize our responses to a continually weirding world. Detailing the transit between affect, culture, and ethics is thus a necessary—albeit thorny—task for environmental thinkers of all stripes.

When we survey these efforts alongside those at once shocking and enervating headlines from the day's news with which I began, we move biomedical and environmental thought out of the silos within which they often reside. We start orchestrating for ourselves that chorus of voices to which Speth opens his ears. This, I believe, is an optimistic practice insofar as it assumes that just responses to environmental and somatic injury are possible, but are only to be found at the intersection of multiple modes of thought and analysis. This practice is also a pedagogical one that has an especially vital home in the humanities classroom. Undoubtedly, just as literature is not a perfect conversion engine, teaching has unpredictable and indeterminable effects. Yet it is a forum for the kinds of experimentation with narrative affect that ecosickness fiction itself undertakes. For this reason, I still heed the words that Charles Altieri appends to *The Particulars of Rapture*: "our most important task as critics, and perhaps as theorists, may be to keep available the possibilities for exclamation built into our affective capacities that are given expression in the arts."[29]

I would add "and perhaps as teachers" to his aside and emphasize that the arts do not only give expression to those affects; they innovatively deploy them. Contemporary fiction is the point of departure for this task in *Ecosickness* because, while inspiring massive action against systemic harms to bodies and planet will require much more than literature and literary analysis alone can accomplish, fiction extends an invitation to read its stories out into the world. It opens channels to the talk between policy and psychology, aesthetics and activism, education and ethics, and data and doxa that positive interventions in pervasive sickness demand. Taking sickness as a crucial organizing category for recent fiction, I have extended this invitation even further. *Ecosickness in Contemporary U.S. Fiction* hopes to encourage readers to venture into the pages of works by Grover, Wojnarowicz, Powers, Wallace, Piercy, and Silko and then to venture beyond them, to find alternative futures at the intersection of the imaginaries and materialities of "today's diseased now."[30]

NOTES

1. ECOSICKNESS

1. Wallace, *Infinite Jest*, 151.
2. Prüss-Üstün and Corvalán, "Preventing Disease Through Healthy Environments," 2.
3. Haynes, *Safe*.
4. Ngai, *Ugly Feelings*, 26.
5. Thacker, *Global Genome*, 61.
6. Alaimo, *Bodily Natures*, 38.
7. Le Sueur, "Eroded Woman," 83.
8. Ibid., 84.
9. Sinclair, *The Jungle*, 30.
10. Ibid., 31.
11. DeLillo, *White Noise*, 107.
12. The subgenre of environmental health memoirs experienced a boom beginning in the 1990s. Titles include Susanne Antonetta, *Body Toxic* (2001); McKay Jenkins, *What's Gotten Into Us?* (2011); Sandra Steingraber, *Living Downstream* (1997); and Terry Tempest Williams, *Refuge* (1991). In *Bodily Natures* Alaimo adeptly analyzes how this corpus of texts brings readers to awareness of the material relatedness between bodies and environments. I reference her arguments throughout this book.
13. Dimock and Wald, "Preface. Literature and Science," 705.
14. Slovic and Slovic, "Numbers and Nerves," 14, 15.

15. For environmental histories of the twentieth century, see Guha, *Environmentalism*; McNeill, *Something New Under the Sun*; Merchant, *American Environmental History*; and Montrie, *A People's History of Environmentalism in the United States*.

16. For a critique of the ethos of "one earth," see Jasanoff, "Heaven and Earth," 31–54. Dipesh Chakrabarty has theorized this planetary consciousness in terms of the Anthropocene, an epoch popularized by Paul Crutzen in which human actions alter not only immediate environments but also geophysical processes. Chakrabarty, "The Climate of History."

17. As Priscilla Wald relates, "the production of recombinant DNA first occurred in a Stanford laboratory in the early 1970s. Stanford applied for a patent on recombinant DNA in the mid-1970s, and it was awarded in 1980, the same year the US Supreme Court confirmed the legality of the first patent on a living organism, a bacterium engineered to break down oil." Wald, "American Studies and the Politics of Life," 200. For more on this history, see Keller, *Refiguring Life*; Kevles, "Out of Genetics," 3–36; and Thacker, *Global Genome*.

18. Rose, *The Politics of Life Itself*, 17–18.

19. Clarke et al., *Biomedicalization*, 47. Part 1 of this edited collection more fully elaborates the shifts to medicalization and biomedicalization. Ivan Illich coined the term *medicalization* in the mid-1970s to disparage the incursion of medicine into all aspects of life—birth, aging, mental distress, discomfort, and death—such that people are unable to manage injury and vulnerability as autonomous individuals. Illich, *Medical Nemesis*, 33.

20. Clarke et al., *Biomedicalization*, 50.

21. Some notes on terminology. My use of the term *biotechnology* is broader than Clarke et al.'s definition: innovations where "basic protein structures are altered affecting genetic structure(s) (e.g., recombinant DNA)." Ibid., 43*n*19. As I use it in this book, *biotechnology* encompasses these inventions as well as those such as pharmaceuticals and diagnostic devices that may—but do not necessarily—use biological agents as the raw components. My use of *technoscience* is more capacious. It captures these technologies and others, such as topical chemical fertilizers, that are not designed to change the matter of life but often do so indirectly. This term also emphasizes that technological development and basic scientific research increasingly go hand in glove after the 1970s. There is a vast literature on the growth of biotechnology, especially as it serves capitalist and state ends. See Cooper, *Life as Surplus*; Rajan, *Biocapital*; Rose and Rose, *Genes, Cells, and Brains*; Smith and Morra, *The Prosthetic Impulse*; and Waldby and Mitchell, *Tissue Economies*.

22. Clarke et al., *Biomedicalization*, 52.

23. The authors of *Biomedicalization* and the other medical sociologists cited here tend to take people in the West as their index cases. It's important to note that, for better and for worse, biomedicine and biotechnology are not equally available to all populations. Chapters 2 and 5 address the differential effects of technology on the poor, women, homosexuals, and the racially and ethnically marginalized.

24. Rose, *The Politics of Life Itself*, 18.

25. See Heise, *Sense of Place and Sense of Planet*.

26. Kleinman, *The Illness Narratives*, 3.

27. Epstein, *Altered Conditions*, 8.

28. I develop this distinction from Mitman, *Breathing Space*, 252.

29. Kleinman, *The Illness Narratives*, 6.

30. See Furberg et al., "Fish Consumption and Plasma Levels."

31. I adapt Ulrich Beck's concept of "risk fate" to the conditions of pervasive sickness. For Beck, a risk fate is a condition "into which one is born, which one cannot escape with any amount of achievement, with the 'small difference' (that is the one with the big effect) that we are *all* confronted similarly by that fate." Beck, *Risk Society*, 41. Whether that fate is one of extreme immiseration or of mild distress depends on a variety of socioeconomic, gender, ethnic/racial, and geographic factors that Beck does not adequately acknowledge in this book.

32. Nixon, *Slow Violence*, 14.

33. Serres, *The Natural Contract*, 33.

34. The literature on ideas of nature is vast, but seminal primers include Cronon, "Introduction" and "The Trouble with Wilderness"; Glacken, *Traces on the Rhodian Shore*; Soper, *What Is Nature?*; and Williams, "Ideas of Nature."

35. Haraway, "The Promises of Monsters," 298.

36. Rabinow, "Artificiality and Enlightenment," 245.

37. Latour, "'It's the Development, Stupid!'"

38. Scott Slovic has usefully defined nature writing as "literary nonfiction that offers scientific scrutiny of the world, . . . explores the private experience of the individual human observer of the world, or reflects upon the political and philosophical implications of the relationships among human beings and the larger planet." Slovic, "Nature Writing," 888.

39. Worster, *Nature's Economy*, 82.

40. Marx, *The Machine in the Garden*.

41. Beck describes reflexive modernization, or its boomerang effect, as occurring when "the agents of modernization themselves are emphatically

caught in the maelstrom of hazards that they unleash and profit from." Beck, *Risk Society*, 37.

42. Lawrence Buell coins "environmentality" to refer to a person's or group's way of "thinking environmental belonging and citizenship." Buell, "Ecoglobalist Affects," 227.

43. I return to this point in the book's conclusion.

44. Altieri, *The Particulars of Rapture*, 8.

45. Leopold, *A Sand County Almanac*, 165.

46. Wolfe, *What Is Posthumanism?* xxv.

47. This strain of posthumanism differs from the techno-utopian variety, which envisions transcending materiality through enhanced integration with computers and other informatic machines.

48. See Alaimo, *Bodily Natures*; Barad, *Meeting the Universe Halfway*; and Bennett, *Vibrant Matter*. See also Wolfe's elaboration of posthumanism as "constitutive dependency and finitude" that deprivileges human consciousness and reason. Wolfe, *What Is Posthumanism?* xxvi. I do not provide a thorough overview of posthumanism and new materialism here. On the various approaches to the former, see Wolfe; and, to the latter, see Coole and Frost, *New Materialisms*.

49. Bennett, *Vibrant Matter*, 9.

50. Alaimo, *Bodily Natures*, 2.

51. Bennett, *Vibrant Matter*, 101–2.

52. Yale University's Forum on Climate Change and Media, with its mission to "analyze and discuss the process by which climate change is communicated through traditional and new media," is one venture that instances this trend. Ward, "NOAA's 2012 Arctic 'Report Card.'"

53. Squier, *Liminal Lives*, 14.

54. Wald, *Contagious*, 3.

55. Ibid., 67.

56. Langston analyzes "the landscape of exposure" to diethylstilbestrol (DES) in postwar America, the hormone disruptor's effects on women's health, and the contests over regulation. Langston, *Toxic Bodies*, xiii. Mitman argues that specific U.S. towns and regions develop in response to allergies and asthma, just as these syndromes themselves result from our meddling in ecosystems. Mitman, *Breathing Space*. Nash shares Langston's and Mitman's focus on the "ecological body," that is, the body "characterized by a constant exchange between inside and outside, by fluxes and flows, and by its close dependence on the surrounding environment." Nash, *Inescapable Ecologies*, 12. She researches a population of

Mexican migrant farmworkers affected by pesticide use in California's Central Valley.

57. Buell, *Writing for an Endangered World*, 31.

58. Alaimo, *Bodily Natures*, 158.

59. Ibid., 156.

60. Nixon, *Slow Violence*, 47. Nixon defines slow violence as "a violence that occurs gradually and out of sight, a violence of delayed destruction that is dispersed across time and space, an attritional violence that is typically not viewed as violence at all." Ibid., 2.

61. Heise's *Sense of Place and Sense of Planet* counsels that we need "an increased emphasis on a sense of planet, a cognitive understanding and affective attachment to the global." Heise, *Sense of Place and Sense of Planet*, 59. Though the book solicits farther-reaching feeling and states that risk is affectively generative, it does not elaborate the various links between narrative, affect, and environmental consciousness. The same observation applies to Buell's essay on "ecoglobalist affects." Its title proclaims an interest in the emotions of global environmentality, but the piece does not differentiate and theorize those affects that constitute what he calls ecoglobalism. Buell, "Ecoglobalist Affects."

Ecocritics Simon C. Estok and Jennifer K. Ladino have focused on specific affects that underpin anthropogenic environmental damage and progressive environmentalism, respectively. Estok, "Theorizing in a Space of Ambivalent Openness" and Ladino, *Reclaiming Nostalgia*. Estok's arguments have been the more contentious of the two. He urges the field to privilege "ecophobia," that is, "irrational and groundless hatred of the natural world, as present and subtle in our daily lives and literature as homophobia and racism and sexism." Estok, "Theorizing in a Space of Ambivalent Openness," 208. While Estok's choice of affect holds a certain logical appeal—that we hate what we fear and therefore destroy it—his affective range is too limited and his theorization of phobia too simplistic. Other literary scholars have elaborated on these complaints. See Taylor, "The Nature of Fear"; and Thornber, *Ecoambiguity*, 9–10.

62. Seminal studies of the pastoral include Alpers, *What Is Pastoral?*; Empson, *Some Versions of Pastoral*; Marx, *The Machine in the Garden*; and Williams, *The Country and the City*.

63. Nancy Easterlin represents another approach to emotion within ecocriticism and is implicitly in dialogue with Wilson. She examines evolutionary psychological theories to elucidate bonds to place and the environmental aesthetics they promote. Easterlin, "'Loving Ourselves Best of All.'" Another important body of work for environmental affect scholarship is

environmental rhetoric. See Harré, Brockmeier, and Mülhaüser, *Greenspeak*; and Herndl and Brown, *Green Culture*.

64. The phrase is Buell's, "Ecoglobalist Affects," 235.

65. In this chapter and throughout this book, I alternate between *affect, emotion,* and *feeling*. Within cultural studies there are no fixed protocols for choosing between these terms. Like many scholars, I do not insist on the distinction. I prefer instead to move between the categories in order to capture the somatic, cognitive, and social dimensions of feelings as well as how they form under historical, cultural, and political pressures. For those interested in debates over the concepts, Brian Massumi provides an influential account of the difference between emotion and affect in Massumi, *Parables for the Virtual*, 28–32. Amélie Rorty thinks historically about the move "from passions to emotions and sentiments." Rorty, "From Passions to Emotions and Sentiments." Altieri differentiates affects, feelings, moods, emotions, and passions in Altieri, *The Particulars of Rapture*, 2. Damasio distinguishes emotion from feeling from a neuroscientific standpoint in Damasio, *Descartes' Error*. Finally, Ngai explains the psychoanalytic bases for distinguishing affect and emotion in Ngai, *Ugly Feelings*, 25–26.

66. Cvetkovich, *Depression*, 4.

67. Lauren Berlant's project on "cruel optimism" draws out *attachment* and *detachment* as the key terms of affect study since the late 1990s. Berlant, *Cruel Optimism*.

68. I borrow this verb from David Palumbo-Liu, who examines the twenty-first-century "'delivery systems'" by which others' affects enter our sphere of concern. Palumbo-Liu, *The Deliverance of Others*, 180.

69. See Clough, "Introduction"; Gregg and Seigworth, "An Inventory of Shimmers"; and Leys, "The Turn to Affect." Gregg and Seigworth usefully differentiate eight, sometimes intersecting, strands of affect study. Gregg and Seigworth, "An Inventory of Shimmers," 6–8.

70. Altieri, *The Particulars of Rapture*, 194.

71. Ibid., 26.

72. Ibid., 194.

73. Ngai, *Ugly Feelings*, 20. See also Brennan, *The Transmission of Affect*.

74. Ahmed, *The Cultural Politics of Emotion*, 6.

75. Altieri, *The Particulars of Rapture*, 263n6.

76. Ibid., 268n19.

77. Berlant, *Cruel Optimism*, 1 (my emphasis).

78. Thornber, *Ecoambiguity*. See especially part 2.

79. Sedgwick, *Touching Feeling*, 128–51.

80. Ngai, *Ugly Feelings*, 5.
81. Quoted in Max, *Every Love Story Is a Ghost Story*, 173.

2. AIDS MEMOIRS OUT OF THE CITY

1. Verghese, *My Own Country*, 14. Hereafter cited parenthetically.

2. Lise Diedrich also notes the similarity between human and microbial mobility. Diedrich, "AIDS and Its Treatments," 241.

3. I use *medical condition* to remain faithful to the distinction between the human immunodeficiency virus, which compromises the immune system, and Acquired Immune Deficiency Syndrome, which HIV causes and which manifests as susceptibility to opportunistic infections. Throughout the rest of this chapter, I follow cultural critics' practice of referring to HIV/AIDS as disease, syndrome, or medical condition.

4. Weston, "Get Thee to a Big City," 255.

5. Ibid., 262.

6. The identity of the alleged but never confirmed "Patient Zero" for HIV crystallizes the mobility of the disease. In 1984, public health officials at the U.S. Centers for Disease Control named Canadian flight attendant Gaëtan Dugas as the index case for HIV infection in North America. Dugas purportedly carried the virus between Europe and North America on his assignments. Randy Shilts's *And the Band Played On* (1987) dramatizes the hunt for Patient Zero and underscores the importance of travel to AIDS research and representation. Shilts, *And the Band Played On*. For a critique of Shilts's book, of Dugas's stigmatization, and of the scientific validity of the Patient Zero concept, see *Zero Patience*, dir. John Greyson, film, 1993; and Wald, *Contagious*.

7. *Modern Nature* (1994), by British filmmaker Derek Jarman, and Eli Clare's *Exile and Pride* (1999) fit neatly within this subgenre of nonurban AIDS memoirs.

8. Halberstam, *In A Queer Time and Place*, 37; Scott Herring, *Another Country*. In addition to Weston and these scholars, anthropologists and socialists have pushed back against the urban focus of queer and AIDS studies. See also Bell and Valentine, "Queer Country"; Dorn and Laws, "Social Theory, Body Politics, and Medical Geography"; Howard, *Men Like That*; Johnson, *Sweet Tea*; and Phillips, Watt, and Shuttleton, *De-Centering Sexualities*.

9. This is only one piece of the epidemiological puzzle. Verghese's research also identifies the phenomenon of "local-locals," residents in the region who acquired the disease near their homes (395). AIDS gains a foothold in this

population through tainted transfusions and sex with infected partners during visits to truck stops. Verghese's paradigm first appeared in the *Journal of Infectious Diseases*. See Verghese, Berk, and Sarubbi, "*Urbs in Rure*."

10. On *My Own Country* as a mode of narrative medicine, see De Moor, "The Doctor's Role of Witness and Companion"; and Diedrich, "AIDS and Its Treatments."

11. Kate Soper provides the terms *nature-endorsing* and *nature-sceptical* to refer to two perspectives on nature: as "discourse-independent" and as a product of cultural construction that "polic[es] social and sexual divisions." Soper, *What Is Nature?* 7.

12. Philip Fisher's telling phrase "privileged settings" refers to "ideal or simplified vanishing points toward which lines of sight and projects of every kind converge," sites that "condense emotional facts." Fisher, *Hard Facts*, 9, 11.

13. Sontag, *Illness as Metaphor and AIDS and Its Metaphors*, 181.

14. Grover, "Constitutional Symptoms," 154.

15. The idea of "crimes against nature" has a legal and not only a discursive force. The accusation was leveled against sodomy, or "buggery," in Europe for centuries before it entered American colonial law in the 1600s. See Eskridge, *Dishonorable Passions*. Verghese comments on a 1986 case in which the district attorney for Johnson City, Tennessee trotted out the state's "crimes against nature" law to prosecute two men discovered having sex. Verghese cites the text of the law: "'Crimes against nature, either with mankind or any beast, are punishable by imprisonment in the penitentiary not less than five years, not more than fifteen.' Included were all acts that did not result in procreation: cunnilingus, fellatio, and anal intercourse" (71). The 2003 U.S. Supreme Court *Lawrence v. Texas* decision proclaimed all state sodomy laws to be unconstitutional. Lawrence v. Texas, 539 U.S. 538 (2003).

16. Simon Watney, "Taking Liberties: An Introduction," in Carter and Watney, *Taking Liberties*, 21.

17. Gould, "The Terrifying Normalcy of AIDS."

18. Treichler, "AIDS, Homophobia, and Biomedical Discourse."

19. Quoted in ibid., 267. Treichler includes background on the biomedical research that led Langone to his conclusions (272).

20. Waldby, *AIDS and the Body Politic*, 20.

21. Ibid., 21.

22. Rotello, *Sexual Ecology*, 1.

23. Ibid., 256–7.

24. Ibid., 189.

25. Ibid., 46.

26. The antinaturalism of much metrocentric AIDS literature and queer theory has an antecedent in Oscar Wilde, whose dialogue "The Decay of Lying: An Observation" (1889) includes the following exclamation: "Enjoy Nature! I am glad to say that I have entirely lost that faculty. People tell us that Art makes us love Nature more than we loved her before; that it reveals her secrets to us. . . . My own experience is that the more we study Art, the less we care for Nature." Wilde, "The Decay of Lying," 3. My thanks to Neville Hoad for directing me to this reference.

27. Soper, *What Is Nature?* 133.

28. Foucault, *The History of Sexuality*, 101. The question whether to appeal to or to eschew genetic reductionism remains germane to early twenty-first-century LGBTQ activism. See, for example, debates surrounding the search for the "gay gene" and Simon LeVay's controversial positions on the brain mechanisms behind sexuality: Brookey, *Reinventing the Male Homosexual*; Hamer, *The Science of Desire*; LeVay, "A Difference in Hypothalamic Structure" and *The Sexual Brain*.

29. Soper, *What is Nature?* 135.

30. Marsh, *Man and Nature*, 11, iii.

31. Botkin, *Discordant Harmonies*, 83.

32. Botkin charts this intellectual history through the writings of Plato, Lucretius, Jean-Jacques Rousseau, Thomas Jefferson, William Wordsworth, Aldo Leopold, and James Lovelock, among others.

33. Plotinus quoted in Botkin, *Discordant Harmonies*, 25. For other efforts to upend classical ecological models, see Barbour, "Ecological Fragmentation in the Fifties"; and O'Neill, "Is It Time to Bury the Ecosystem Concept?"

34. A deep anthropocentrism can also undergird claims that earth is a harmonious system insofar as it is the sustenance of *human* life that determines an ecosystem's health. This is known as the "anthropic principle," the idea "that the fundamental laws of the universe are 'tuned' to permit the evolution of life and consciousness." Botkin, *Discordant Harmonies*, 213*n*3.

35. Ibid., 62.

36. Phillips, *The Truth of Ecology*. As other ecocritics have remarked, Phillips's account of ecocriticism and of realism's place within it can be reductive. That said, he drew needed attention to how scientific models circulate as metaphorical truths within environmental writing. See Buell, *The Future of Environmental Criticism*; Heise, "The Hitchhiker's Guide to Ecocriticism"; and Gifford, "Recent Critiques of Ecocriticism." I take up Phillips's claims again in chapter 3.

37. Social theorists of medicine are instructing today that somatic health is like Botkin's "static landscape": it never existed except in our imaginings of it. Writing "against health," Jonathan Metzl cites Ivan Illich's declaration that "health" "'is the most cherished and destructive certitude of the modern world.'" Quoted in Metzl and Kirkland, *Against Health*, 5. "Destructive" because it turns pain and impairment into unnatural states based on clinical measures calibrated to generalized norms of somatic harmony. But it is health, these critics contend, that is the unnatural state, one that a person can only approach asymptotically and with the aid of technological interventions. Just as the "discordant harmony" about which Botkin writes is difficult to swallow because it challenges inherited norms, it is difficult to accept that the body is a stochastic system within which failure is normal (sensu Perrow). Perrow, *Normal Accidents*.

38. Greg Garrard has recently questioned the satisfactions of the category of health in environmental writing. He objects to "life writing about disease and death" that uses health as a normative value for gauging environmentalist and ecocritical projects. Garrard, "Nature Cures?" 494.

39. I am not discrediting the ecopolitical potential of aesthetic appreciation. Samuel Hays has shown that, as Americans accumulated discretionary income and leisure, they began spending this capital on outdoor recreation and travel. They thereby came to value and to protect environmental "amenities." Hays, *Beauty, Health, and Permanence*, 4. These amenities had to have a particular character, however. Attracted by the majestic beauty and seeming permanence of the wilderness, nature photographers helped establish the picturesque and the healthy as constitutive of pristine nature (37).

40. Grover, *North Enough*, 7. Hereafter cited parenthetically.

41. It is worth noting that Grover's and Wojnarowicz's journeys are solitary. They are not interested in reviving utopian gay separatist communities.

42. Cvetkovich, *An Archive of Feelings*, 211.

43. Liz Bury coined the term in Bury, "Tugging at Heart Strings." Paradigmatic examples of each of these categories include Mary Karr, *Lit: A Memoir* (2009); Kathryn Harrison, *The Kiss* (1997); Joyce Carol Oates, *A Widow's Story: A Memoir* (2011); and Jean-Dominique Bauby, *Le scaphandre et le papillon* (*The Diving Bell and the Butterfly*; 1997).

44. Cvetkovich, *An Archive of Feelings*, 218.

45. "Still point" alludes to the "Burnt Norton" section (first published in 1936) of T. S. Eliot's *Four Quartets* (1943), lines from which form the epigraph to the memoir's second section, "Cutover": "*To arrive where you are, to get*

from where you are not, / You must go by a way wherein there is no ecstasy. / In order to arrive at what you do not know / You must go by a way which is the way of ignorance." Grover, *North Enough,* 9.

46. Thoreau, *Journal,* 351.

47. Just as AIDS suffuses Grover's environmental perception, it constructs her relation to time. Throughout the memoir, she marks time using friends' dates of death. For example, "Perry died in September 1991. The following spring, I impulsively decided to buy a piece of the Minnesota dream: a cabin in the north woods" (5) and "the fall after James died, I took the train to Hudson Bay" (148). Like the AIDS artists whom Lauren Berlant studies, Grover inhabits *"crisis time"*: she "lives the present intensely," but she does so through environmental immersion. Berlant, *Cruel Optimism,* 59.

48. Ahmed, *The Cultural Politics of Emotion,* 147.

49. It is worth noting that the memoir makes subtle comment on standards of gay male beauty as well as environmental beauty. Grover suggests that the body aesthetic dominant in San Francisco's Castro district exacerbated the sense of self-loss that agonized gay men with AIDS. See the vignette featuring Eric (57–70).

50. Fetterley and Pryse, *Writing Out of Place,* 315, 316.

51. Ibid., 33.

52. Verghese, *My Own Country,* 22, 294.

53. Botkin, *Discordant Harmonies,* 62.

54. Cronon, "The Trouble with Wilderness," 89.

55. Mortimer-Sandilands, "'I Still Need the Revolution,'" 69.

56. Wojnarowicz, *Close to the Knives,* 26. Hereafter cited parenthetically.

57. Alaimo, *Bodily Natures,* 87. At the same time, because Wojnarowicz recounts the sexual and physical abuse he received as a child and his years in New York scoring drugs and funding his habit through prostitution, *Close to the Knives* also fits the subgenre of "misery" memoirs.

58. The political power of *Delta Towels* and other works by Wojnarowicz is clear from their use in antiobscenity campaigns. In 1990 Wojnarowicz sued the American Family Association under New York's Artists' Authorship Rights Act because the group reproduced his art in fund-raising materials. According to court documents, the American Family Association's mission was to "promot[e] decency in the American society and advanc[e] the Judeo-Christian ethic in America." Wojnarowicz v. American Family Association, at 1. The U.S. District Court of New York ruled in Wojnarowicz's favor.

59. Johnson, *Sweet Tea,* 5.

60. Reviewer David Finkle notes the pitch of paranoia in this frequently repeated phrase and in the memoir's overall tone. Finkle, "Postcards from America," 239.

61. For a history of these drug therapies, see Zuniga et al., *A Decade of HAART*. For an account of antiretroviral therapies in AIDS treatment policies in the developing world, see Biehl, *Will to Live*.

62. Wojnarowicz was not HIV positive for all of the events that *Close to the Knives* relates. When I refer to Wojnarowicz as "sick," I suggest either the seropositive status that came to light in the mid-1980s or to his status as lover, friend, and caregiver to other PWAs.

63. Todd R. Ramlow examines borderlands and borders such as fences in *Close to the Knives* as occasions for "an ongoing and mobile revisioning of subjectivity" that dismantles binarisms separating disabled from healthy and queer from hetero. Ramlow, "Bodies in the Borderlands," 171.

64. Mumford, *The City in History*, 494.

65. The "universe"-al phrases that pepper *Close to the Knives* can grate on the reader. These phrases dichotomize "us and them" in a rhetoric that sounds too similar to suburbanites' ideological and physical "subdividing." Talk of "universes" and the "one-tribe nation" stifle nuance, but they enhance the memoir's polemic against structural oppression and hate (37).

66. Smith, *Uneven Development*, 33.

67. In *Close to the Knives,* technomorphism updates the trope of body-environment merging. Elsewhere, the narrator ponders the body rendered electronic—"What do these eyes have to do with surveillance cameras? What do the veins running through my wrists have in common with electric wiring?" (63)—and decides that cyborgism is essential to his artistic project: "I'm the robotic kid with caucasian [sic] kid programming trying to short-circuit the sensory disks. I'm the robotic kid looking through digital eyes" (63). James Romberger's illustrations for *Seven Miles a Second* (1996), a comic book roughly based on Wojnarowicz's memoir, nicely translate the book's interwoven organic and inorganic imagery. Wojnarowicz and Romberger, *Seven Miles a Second*.

68. The fluidity between Wojnarowicz's identity and the prisoner's, established through the repetition of "watery" in the first two sentences, indicates another of the memoir's habits. Primarily through grammatical ambiguity, the narrator conflates his body parts with those of lovers and passersby (29), a city's architecture with the bodies that inhabit it (32, 35), and a city with the disease that devastates it (30). In *Seven Miles a Second*, Wojnarowicz makes his desire to merge with another explicit: "If I could attach our blood vessels

so we could become each other, I would. If I could attach our blood vessels in order to anchor you to the earth to this present time, I would." Wojnarowicz and Romberger, *Seven Miles a Second*, 55.

69. Sontag, *Illness as Metaphor*, 3.

70. Altieri, *The Particulars of Rapture*, 209.

71. Ibid., 188. Altieri's "intensity" is not cognate with Brian Massumi's "intensity," which is equivalent to autonomic affect. See Massumi, *Parables for the Virtual*, 23–28.

72. Fisher, *The Vehement Passions*.

73. Alaimo, *Bodily Natures*, 89, 95.

74. By contrast, the "I" of *My Own Country* is secure throughout, and Verghese's memoir reads as the *bildung* of a rising doctor and new American.

75. Two groups of scholars join this phalanx: those who are rethinking depth hermeneutics and those who are interested in how interpretive modes inflect or even impinge on life practices. For the former efforts, see Best and Marcus, "Surface Reading"; and Moretti, *Graphs, Maps, Trees*. For the latter, see Anderson, *The Way We Argue Now*; and Love, "Close but not Deep."

76. Ricoeur, *Freud and Philosophy*, 33.

77. Sedgwick, *Touching Feeling*, 124.

78. Felski, "Suspicious Minds," 219.

79. The next chapter elaborates on Sedgwick's claims with respect to Richard Powers's *The Echo Maker*.

80. Cheng, "Skins, Tattoos, and Susceptibility," 100–1. Latour's cognate proposal is derived from his reading of Alan Turing: "What would critique do if it could be associated with *more*, not with *less*, with *multiplication*, not *subtraction*. Critical theory died away long ago; can we become critical again, in the sense here offered by Turing? That is, generating more ideas than we have received, inheriting from a prestigious critical tradition but not letting it die away." Latour, "Why Has Critique Run Out of Steam?" 248.

81. Felski, "After Suspicion," 31. Felski's affective and political claims about suspicion become confusing as they multiply. Is suspicion bad because it blocks openness and pleasure? Or is it bad because it does not recognize that "it offers its own substantive pleasures?" Felski, "Suspicious Minds," 228. Is suspicion bad because, in guarding himself against otherness, the critic shuts down a positive reading "that is equipped with momentous political implications?" Felski, *Uses of Literature*, 5. Or is it simply tired and banal and therefore politically impotent? Felski, "Suspicious Minds," 231. As Felski's suspicion is all of these things at once, her critique loses its teeth.

82. Felski, "Suspicious Minds," 219–20.

83. Felski, *Uses of Literature*, 14.
84. Sedgwick, *Touching Feeling*, 124.
85. Adorno, *Negative Dialectics*, 203.
86. Morton, *Ecology Without Nature*, 24.
87. Ibid.
88. Nixon, *Slow Violence*, 143.
89. Though environmental justice contests had been fought for decades, principles for the movement were first codified in 1991 at the First National People of Color Environmental Leadership Summit in Washington, DC. For more on this history and the voices of the movement, see the first three chapters of Adamson, Evans, and Stein, *The Environmental Justice Reader*.
90. Like the material memoirists that Alaimo examines, Grover and Wojnarowicz validate firsthand experience. Unlike Alaimo's authors, however, they do not necessarily incorporate hard scientific data in order to do so. Alaimo, *Bodily Natures*, 87.
91. Epstein, *Impure Science*, 12.
92. Ibid., 14.
93. This tension repeated itself in AIDS activists' and educators' relationship to the media. With or without malicious intent, newspapers, popular magazines, scholarly journals, and TV programs propagated inaccurate or incomplete information. Cindy Patton reports that "media science frequently articulated pre-existing stereotypes in a new, objective-sounding language. . . . While the media have been instrumental in raising social and medical awareness about AIDS, the reportage has consistently misrepresented the basic concepts of HIV, sensationalized faulty research, and selectively reported on conflicting data." Patton, *Inventing AIDS*, 26–27. Even as they sought media attention, then, AIDS activists urged the public to take seriously frictions between what they read and what they felt of the disease.
94. International Covenant on Economic, Social, and Cultural Rights, United Nations.
95. Nietzsche, *The Gay Science*, 177.
96. Epstein, *Impure Science*, 336. See the following on how social health movements based challenges to medical consensus on individuals' illness experience: Brown, *Toxic Exposures*; Corburn, *Street Science*; and Kroll-Smith and Floyd, *Bodies in Protest*.
97. Epstein, *Impure Science*, 337.
98. Di Chiro, "Local Actions, Global Visions," 209.

99. Catriona Mortimer-Sandilands and Bruce Erickson, "Introduction," in Mortimer-Sandilands and Erickson, *Queer Ecologies*, 30.

100. Womack, "Suspicioning," 145.

101. Soper, *What Is Nature?* 16.

3. RICHARD POWERS'S STRANGE WONDER

1. Quoted in Daston and Park, *Wonders and the Order of Nature*, 303.

2. Dawkins, *Unweaving the Rainbow*, xii.

3. For classic and latter-day accounts of the pastoral, see Bate, *Romantic Ecology*; Empson, *Some Versions of Pastoral*; Garrard, *Ecocriticism*; Gifford, *Pastoral*; Marx, *The Machine in the Garden*; and Williams, *The Country and the City*.

4. Yoon, "Luminous 3-D Jungle Is a Biologist's Dream."

5. Ibid.

6. Descartes, *The Passions of the Soul*, 56.

7. International Union for the Conservation of Nature, "Summary Statistics for Globally Threatened Species."

8. Ehrlich, *The Population Bomb*, 56–57.

9. Smith, "Postmodernism and the Affective Turn." Stephen Burn notes Powers's "concern for the planet," and elaborates that "Powers's work is motivated by an ecological vision that tries to probe the glistening veneer of contemporary media reality . . . to reach the axis of interconnected life that lies beneath." Powers, "An Interview with Richard Powers," 164. Joseph Tabbi uses environmental metaphors to characterize the effect of Powers's postmodernism: "his fiction is ecological in a wider sense, opening connective possibilities through disciplinary knowledge of the cultural environment, and written in a recursive language and self-reflexive style that produces the literary equivalent of an ecosystem." Tabbi, *Cognitive Fictions*, 61. For a similar tendency to describe Powers's texts through metaphors of ecology, see Atwood, "In the Heart of the Heartland." Confirming his growing concern for ecological dilemmas, Powers participated in a symposium on environmental representation at Stanford University. Powers, "Environmental Writing in Four Dimensions: Fiction."

10. Hale, "Aesthetics and the New Ethics," 901.

11. Powers, *The Gold Bug Variations*, 592.

12. Powers, *The Echo Maker*, 359. Hereafter cited parenthetically.

13. Charles Harris coins the phrase *neurological realism* to distinguish Powers's brand of realism from psychological realism à la Henry James and

Virginia Woolf. My analysis follows Harris's argument that Powers's novels "foreground the effects of largely unconscious neurological activities" and "dismantle . . . dualisms on neuroscientific grounds." Harris, "The Story of the Self," 243-44.

14. Thomashow, *Bringing the Biosphere Home*, 57.

15. Ibid., 67.

16. Carson, *The Sense of Wonder*, 42–43.

17. Abbey, *Desert Solitaire*, 36-37.

18. The tension between the pressures of fact and of feeling links Powers's effort to that of the AIDS memoirists examined in chapter 2.

19. Powers, "The Art of Fiction," 131.

20. Powers, *The Gold Bug Variations*, 611.

21. Ibid., 325, 8.

22. Shklovsky, "Art as Technique," 12.

23. Ibid., 18. Philosopher John Dewey's claims for art echo Shklovsky's and illuminate Karin's wonder-filled experience: "Familiarity induces indifference. . . . Art throws off the covers that hide the expressiveness of things; it quickens us from the slackness of routine and enables us to forget ourselves by finding ourselves in the delight of experiencing the world about us in its varied qualities and forms." Dewey, *Art as Experience*, 108.

24. *The Echo Maker* alludes to its own project later in the narrative. On one of her house visits, Mark's rehabilitation nurse, Barbara, captivates Mark with an art history book entitled *A Guide to Unseeing: 100 Artists Who Gave Us New Eyes* (241).

25. Powers, "Making the Rounds," 306.

26. Descartes, *The Passions of the Soul*, 52.

27. Izard and Ackerman, "Motivational, Organizational, and Regulatory Functions of Discrete Emotions," 257.

28. For Sara Ahmed, wonder is a historical and embodied relation because it accompanies new sight. It is "about learning to see the world as something that does not have to be, and as something that came to be, over time, and with work." Ahmed, *The Cultural Politics of Emotion*, 180. *The Echo Maker* displays this dimension of wonder when it elaborates on the history of agricultural and tourist development in Buffalo County, particularly through references to Willa Cather's *My Ántonia* (1918).

29. Fuller, *Wonder*, 101.

30. The oscillation between exposition and lyricism is different from the oscillation that Powers notes in a 2008 essay on his prose. He explains that "the novel I'm after functions as a kind of bastard hybrid, like consciousness

itself, generating new terrain by passing 'realism' and 'metafiction' through relational processes, inviting identification at one gauge while complicating it at others." Powers, "Making the Rounds," 308. One can argue that *The Echo Maker* performs this operation through the postmodern tactics such as subjective inconsistency that critics such as Rachel Greenwald Smith have identified. Smith, "Postmodernism and the Affective Turn," 424. But what interests me is how "new terrain" builds up when exposition and lyricism cooperate to generate wonder.

31. Quoted in Harris, "The Story of the Self," 238. For a detailed analysis of the novel's focalization techniques, see Herman and Varvaeck, "Capturing Capgras."

32. This does not mean that *The Echo Maker* portrays extant research on human neurobiology as exhaustive, only that the narrator exhibits greater confidence when communicating insights from neuroscience.

33. Powers, *The Gold Bug Variations*, 592.

34. Buell, *The Environmental Imagination*, 92.

35. Ibid., 102.

36. Ibid., 104.

37. This description sets the precedent for portraying Mark using zoomorphism and technomorphism, a technique that I will analyze further on.

38. Fisher, *Wonder, the Rainbow, and the Aesthetics of Rare Experiences*, 7. For Fisher, only visual phenomena elicit wonder because one can take in both the whole and its details at once. Powers's work suggests that he errs on this point. Nonsynchronous narrative arts perhaps best produce the "slow unfolding of attention" that wonder demands, as the recursivity of reading permits one to grasp details in a whole. Ibid., 6. As the next chapter will argue, this aspect of narrative is also crucial to the mechanics of disgust in David Foster Wallace's *Infinite Jest*.

39. Ibid., 73.

40. Foucault, "The Masked Philosopher," 328.

41. *The Echo Maker*'s account of the intermingling of familiarity and strangeness develops from Powers's 1995 novel, *Galatea 2.2*. In this text, researcher Ram Gupta marvels at the precariousness of consciousness by adducing the case of prosopagnosics. These are people who can no longer recognize friends, celebrities, and even themselves due to brain injury. In an exchange with the protagonist, named Richard Powers, Ram remarks, "'You know, I think the astonishing may be the ordinary by another name. But these results do lead us to many tempting guesses.... That everything you

are capable of doing could be taken away from you, in discrete detail.' [Richard] added to Ram's list the obvious, the missing speculation. . . . That what you loved could go foreign without your ever knowing. That the eye could continue tracing familiarity, well into thought's unknown regions." Powers, *Galatea 2.2*, 299. Whereas *Galatea 2.2* only speculates on this scenario, *The Echo Maker* examines it in situ.

42. Fuller, *Wonder*, 101.

43. Hirstein and Ramachandran, "Capgras Syndrome," 437.

44. Feinberg and Keenan, "Where in the Brain Is the Self?" 667.

45. Powers, *The Echo Maker*, 102. For a history of how psychologists and neuroscientists have classified Capgras and how those classifications inform *The Echo Maker*, see Draaisma, "Echos, Doubles, and Delusions."

46. Didactic moments such as those that I'm highlighting here do not appear in *The Echo Maker* as a continuous, sustained lesson. Rather, events in Weber's life trigger exposition or, as in this case, the narrative delivers information without events obviously motivating it.

47. Later in the story, Karin references a *People's Free Dictionary* entry for Frégoli syndrome that repeats the view that mental health and dysfunction are on a continuum. "Some researchers," it reads, "suggest that all misidentification delusions may exist along a spectrum of familiar anomalies shared by ordinary, nonpathological consciousness" (261).

48. Smith also identifies oscillations between "recognizable and unrecognizable forms of consciousness" in *The Echo Maker* and maps this oscillation onto the novel's dual commitments to "postmodernist distance . . . and psychological realism's illusion of mimesis." Smith, "Postmodernism and the Affective Turn," 432, 433. She reads these oscillations as evidence that postmodernist techniques produce autonomic, pre-subjective affectivity.

49. The definition of place that emerges from *The Echo Maker*'s storyworld accords with Buell's concise definition of it: "space that is bounded and marked as humanly meaningful through personal attachment, social relations, and physiographic distinctiveness. Placeness, then, is co-constituted environmentally, socially, and phenomenologically through acts of perception." Buell, *The Future of Environmental Criticism*, 145. As I use it, *environment* is both broader than *place* and more active than perceptual. That is, in ecosickness fiction, the environment is not only a place that characters perceive and inhabit and that exerts pressure on their identities; it also elicits intervention in the form of manipulation and exploitation as well as concern, restoration, and protection.

50. Augé, *Non-Places*, 160.

51. The epigraph to part 1, which directly precedes this passage, contextualizes the use of the present tense. It reads, "We are all potential fossils still carrying within our bodies the crudities of former existences, the marks of a world in which living creatures flow with little more consistency than clouds from age to age" (1). Extracted from Nebraskan Loren Eiseley's *The Immense Journey* (1957), which examines the traces of evolution in his surroundings, this line fuses past ("former existences"), present ("still carrying"), and future ("potential fossils") and foreshadows the narrative's deep account of crane behavior. Eiseley, *The Immense Journey*.

52. I use a less restricted meaning of geographical determinism than is current with geographers, historians, and anthropologists who use the phrase to mean that geography and climate have influenced the rise and fall of societies. I refer more broadly to how a place can limit or expand an individual's life possibilities.

53. Marx, *The Machine in the Garden*, 109–10.

54. de Crèvecoeur, *Letters from an American Farmer*, 47.

55. Iowa Beef Processors was one of the largest meat processing and packing corporations when Tyson Fresh Meats purchased it in 2001. In the novel, Mark works at the Lexington, Nebraska plant, which also figured in Eric Schlosser's investigation of its slaughtering practices in *Fast Food Nation* (2001).

56. With Cather's *My Ántonia* as an explicit intertext, *The Echo Maker* contributes to a literary tradition of deromanticizing Plains settlement through mental illness while simultaneously exalting the landscape for its wondrousness. Relating Mr. Shimerda's failure to learn viable farming practices and his eventual suicide, Cather's novel proves that the land can produce a bad sort of "flavor" just as often as a pleasant one. Writing for the *Atlantic Monthly* in 1893, E. V. Smalley contextualizes Cather's story of the Homestead Act's failure to deliver economic opportunity and well-being to those in marginal populations: "an alarming amount of insanity occurs in the new prairie States among farmers and their wives. In proportion to their numbers, the Scandinavian settlers furnish the largest contingent to the asylums." Smalley, "The Isolation of Life on Prairie Farms," 380.

57. See Mittal and Kawaai, "Freedom to Trade?"

58. Thomashow, *Bringing the Biosphere Home*, 57.

59. Ibid., 212.

60. Cameron, *Writing Nature*, 44.

61. Powers, "The Last Generalist."

62. Powers, "An Interview with Richard Powers," 110.

63. For more on how vulnerability is built into complexity see Hayles, *How We Became Posthuman*; and Perrow, *Normal Accidents*.

64. Hacking, "Our Neo-Cartesian Bodies in Parts," 80.

65. Zunshine, *Strange Concepts*, 55.

66. Defining Weber as a holist may sound contradictory given his rebuttal of the idea that we are "one, continuous, indivisible whole" (171). This is not inconsistent; the keywords in Weber's statement are "continuous" and "indivisible."

67. Immediately following these musings on the role of stories in neurology, Weber and Mark discuss xenotransplantation: "the growing body of experiments [in which] bits of cortex from one animal [are] transplanted into another, taking on the properties of the host area" (416). The juxtaposition of these scenes highlights the contrast between holistic approaches to treatment and surgical, functionalist approaches.

68. Rizzolatti and Sinigaglia, *Mirrors in the Brain*, xii. *The Echo Maker* retells the experiments that led the researchers to their conclusions:

> Every time the monkey [in Rizzolatti's study] moved its arm, the neurons fired. One day, between measurements, the monkey's arm neurons began firing like crazy, even though the monkey was perfectly still. More testing produced the mind-boggling conclusion: the motor neurons fired when one of the lab experimenters moved *his* arm. Neurons used to move a limb fired away simply because the monkey saw *another* creature moving, and moved its own imaginary arm in symbol-space sympathy. (355)

69. Rizzolatti and Sinigaglia, *Mirrors in the Brain*, xii.

70. Ibid. For an overview of research on the relationship between empathy and mirror neurons, see Gallese, "'Being Like Me'"; and Iacobini, *Mirroring People*. Over the past decade, literary scholars have looked to cognitive approaches and models to rethink narrative's capacity to activate or impede identification. See, especially, Boyd, *On the Origin of Stories*; Keen, *Empathy and the Novel*; and Zunshine, *Strange Concepts*.

71. Ivan Illich adduces the phenomenon of iatrogenesis in his critique of medicalization. He defines iatrogenesis as "any adverse condition in a patient occurring as the result of treatment by a physician or surgeon." Illich, *Medical Nemesis*, 14.

72. Rizzolatti and Sinigaglia, *Mirrors in the Brain*, xii.

73. In its last pages, *The Echo Maker* reveals that Barbara caused Mark's accident in her botched suicide attempt: she put herself in front of Mark's speeding car on North Line Road. In part 1, however, Karin believes Barba-

ra's attentive care for Mark is selfless and not her surreptitious way of atoning for her actions.

74. Commoner, *The Closing Circle*, 33. John Muir's oft-cited reflection also resonates here: "When we try to pick out anything by itself, we find it hitched to everything else in the universe." Muir, *My First Summer in the Sierra*, 211.

75. Phillips, *The Truth of Ecology*, 45.

76. Ibid., 51.

77. The following works also trace the shift away from models of ecological harmony and equilibrium: Barbour, "Ecological Fragmentation in the Fifties"; Botkin, *Discordant Harmonies*; O'Neill, "Is it Time to Bury the Ecosystem Concept?"; and Worster, *Nature's Economy*.

78. Sedgwick, *Touching Feeling*, 126.

79. Brennan, *The Transmission of Affect*, 3.

80. The paranoiac's quest to expose hidden plots, another feature that Sedgwick outlines, is also relevant to *The Echo Maker*. The novel is, in part, a detective story. An indecipherable clue to the cause of Mark's accident—a note scrawled in a "spidery, ethereal" hand—guides the plot (10). It appears at Mark's bedside in the trauma unit and delivers this message: "I am No One / but Tonight on North Line Road / GOD led me to you / so You could Live / and bring back someone else" (10). Readers learn that Barbara is the addressee of these verses and Mark its author. He penned the note, which anticipates the novel's outcome, before he slipped into his coma.

81. The following reviews represent the various positions on Powers's preoccupation with complexity: Deresiewicz, "Science Fiction"; Michiko Kakutani, "Imaginary Lives, Built in Empty Rooms"; and Sutherland, "Paper or Plastic?"

82. Powers, *The Gold Bug Variations*, 611.

83. Ibid., 336.

84. Ibid.

85. Ibid. With different motives and outcomes, Jan raises the questions about the ethical implications of technoscientific change that motivate Leslie Marmon Silko's *Almanac of the Dead* and that I analyze in chapter 5.

86. Ibid., 387.

87. Rose, *The Politics of Life Itself*, 24, 146.

88. Sontag, *Illness as Metaphor and AIDS and Its Metaphors*, 3.

89. Haraway, *Simians, Cyborgs, and Women*, 150.

90. Powers, *The Gold Bug Variations*, 245.

91. Claudine Herzlich places this principle in a historical context, arguing that "medicine long failed to recognize the psychological and social

250   3. RICHARD POWERS'S STRANGE WONDER

consequences of labelling illness and felt that while diagnosis and treatment could sometimes be of no use, they could never be harmful. Hence the medical rule that it is better to mistakenly diagnose a healthy man as sick than a sick man as healthy." Herzlich, "Modern Medicine and the Quest for Meaning," 159.

92. This tenet appears in book 1 of Hippocrates's *Epidemics*.

93. As Karin's position relative to environmental endangerment evolves, her position on caring for Mark changes as well. She no longer makes him her life's project nor defines his health in her own terms. "The goals of care had changed," the narrator announces. "She no longer needed him to recognize her. She only needed him to believe he was alive" (398).

94. Harrison, *Gardens*, 25.

95. Even Daniel admits, "'those birds are doomed'" (56).

96. This book's conclusion takes up the possibilities for optimism in a sick world.

97. Deresiewicz disagrees with the first part of this claim. He chastises Powers for not "letting the story speak," and argues that Daniel ventriloquizes Powers's principles, leaving no room for readers' objections. Deresiewicz, "Science Fiction," 28. "The novelist who refuses to grant his readers imaginative and moral freedom—the two are the same, and connected to the characters' own autonomy—is serving neither the cause of art nor of justice," he pronounces. Ibid. Deresiewicz misses one of the novel's central projects, however: to wrangle with the meaning of autonomy in the face of the complex connectedness that neural and environmental damage bring to light.

98. Caldwell, "Awakenings."

99. Slovic, *Seeking Awareness in American Nature Writing*, 14.

4. *INFINITE JEST*'S ENVIRONMENTAL CASE FOR DISGUST

1. See chapter 1n16.

2. Buell, *The Future of Environmental Criticism*, 46.

3. "Plane Stupid." Credit for the rise of "subvertising," advertising that aims to subvert the paradigms of growth and consumption, goes to Canadian magazine *Adbusters*.

4. Other frames from the video call on viewers' memories of the iconic and much debated image of the "falling man" who jumped out of the burning North Tower of the World Trade Center on 11 September 2001. This evocation is clearly deliberate, given that the group is targeting air travel and that planes were the weapons in Al Qaeda's plot.

5. Morton, *Ecology Without Nature*, 159.
6. Rozin, Haidt, and McCauley, "Disgust," 757.
7. Pole, "Disgust and Other Forms of Aversion," 228, 226.
8. Wallace, "Interview with David Foster Wallace," 131.
9. Ibid., 132. Letter to Michael Pietsch dated 22 June 1992, quoted in Max, *Every Love Story Is a Ghost Story*, 173.
10. Hobbes, *Leviathan*, 38.
11. The text deliberately muddles the translation between "Subsidized Time" and the Gregorian calendar; these are best guesses.
12. The play in this acronym already points to one of *Infinite Jest*'s targets: the many forms of self-absorption that plague contemporary America.
13. Wallace, *Infinite Jest*, 666. Hereafter cited parenthetically.
14. Wallace, "Interview with David Foster Wallace," 127.
15. David Foster Wallace, "E Unibus Pluram: Television and U.S. Fiction" (1993) in *A Supposedly Fun Thing I'll Never Do Again*, 37. Hereafter cited parenthetically.
16. The intelligence community has another name for "Infinite Jest": the *samizdat*. Hal defines this insurgent cultural form as "'the sub-rosa dissemination of politically charged materials that were banned when the . . . Kremlin was going around banning things. Connotatively, the generic meaning now is any sort of politically underground or beyond-the-pale press or the stuff published thereby" (1011*n*110).
17. In "Infinite Jest"'s other scene, Joelle repeatedly walks through a revolving door until she glimpses someone familiar revolving in the opposite direction. They continue to revolve, pursuing each other, for several rotations. This is only one account of the film's content. The variety of inconsistent accounts and the contexts of interrogation in which they emerge make the actual content uncertain. See Molly Notkin's version on 788f.
18. Wallace, "An Interview with David Foster Wallace," 127.
19. In *Infinite Jest*, "E Unibus Pluram" is O.N.A.N.'s ironic motto.
20. Wallace would undoubtedly agree with art historian Rosalind Krauss when she argues that "video's real medium is a psychological situation, the very terms of which are to withdraw attention from an external object—an Other—and invest it in the Self," quoted in Lev Manovich, "Database as Symbolic Form," 52.
21. David Foster Wallace, "Joseph Frank's Dostoevsky" (1996) in *Consider the Lobster and Other Essays*, 261.
22. Wallace, "Interview with David Foster Wallace," 150.
23. "A Failed Entertainment" was Wallace's original title for *Infinite Jest*.

24. Wallace's complex sentence structures and diction direct attention to the author's grammatical and lexical choices. Similarly, the details of Wallace's biography encourage us to examine the minutiae of his writing. Wallace's mother, Sally Foster, authored a grammar primer entitled *Practically Painless English* and inspired Avril Incandenza's profession in *Infinite Jest*. Avril, who succeeds her husband as dean of E.T.A., is a strict grammarian who polices her charges' English. Wallace addresses questions of grammar and usage directly in his 2001 *Harper's* essay, "Tense Present: Democracy, English, and the Wars over Usage," which appears as "Authority and American Usage" in *Consider the Lobster*.

25. Oatley, *Best Laid Schemes*, 56.

26. Feagin, "The Pleasures of Tragedy," 97.

27. Wallace, *Brief Interviews with Hideous Men*, 59.

28. Jameson, *Postmodernism*, 48–49.

29. Wallace does not, however, celebrate art's capitulation to capitalist dictates of profit and entertainment. As "E Unibus Pluram" suggests, he seeks an emotionally—if not necessarily politically—redemptive fiction, but the trope of "distance" is a precarious one for Wallace.

30. Packard, *The Waste Makers*, 6.

31. Ibid., 160.

32. Ibid., 238.

33. Hayles, "The Illusion of Autonomy," 685.

34. The policy of distancing also links the novel's waste and tennis plots. One of Jim Incandenza's films explains that the goal of tennis is precisely to send away a ball that you hope your opponent will not return.

35. By no means is *Infinite Jest*'s United States pollution-free, however. While many features distinguish Boston's affluent neighborhoods from the squalid ones, trash is a great equalizer. As Don Gately cruises from Enfield to Cambridge, the narrator inventories the abundant litter: "an odd little tornado of discarded ad-leaflets and glassine bags and corporate-snack bags and a syringe's husk and filterless gasper [i.e., cigarette]-butts and general crud and a flattened Millennial Fizzy cup" (479). This "tornado of waste" ruining Boston's streets in fact forms the city's bedrock (479). Litter "eventually becomes part of the composition of the street. A box is like intaglioed into the concrete of the sidewalk" (583). Together, these descriptions of pollution prove the failure of C.U.S.P.'s agenda for a "Tighter, Tidier Nation."

36. It is worth noting that Mario's is only one reconstruction of the birth of O.N.A.N. The narrator makes it clear that the show is a parody of his father's parody of the historical events (see 400 and 438).

37. Wallace, *The Broom of the System*, 53.

38. "KAB: A Beautiful History," Keep America Beautiful, accessed 26 March 2013, http://www.kab.org/site/PageServer?pagename=kab_history. TV viewers know this group from its iconic and much criticized "Crying Indian" advertising campaign from the 1970s. The novel alludes to this campaign when Orin Incandenza reminisces about the broadcast era of TV: "'I miss late-night anthems and shots of flags and fighter jets and leathery-faced Indian chiefs crying at litter'" (599).

39. Packard, *The Waste Makers*, 238.

40. Nash, *Inescapable Ecologies*, 7.

41. Ibid., 12.

42. Ibid., 11.

43. Alaimo, *Bodily Natures*, 2, 15.

44. Elizabeth Freudenthal shares my interest in the biomedical imagination of the novel but focuses on how it produces an "anti-interior selfhood" as characters exteriorize identity through compulsive relations with the object world. Freudenthal, "Anti-Interiority," 193.

45. Hayles, "The Illusion of Autonomy," 676.

46. Ibid., 695.

47. The ATHSCME company produces fans for blowing waste over the U.S.-Canada border.

48. To put this in geographer Neil Smith's terms, "By producing the means to satisfy their needs, human beings collectively produce their own material life, and in the process produce new human needs whose satisfaction requires further productive activity" and, I add, further damage. Smith, *Uneven Development*, 55.

49. Buell, *Writing for an Endangered World*, 31.

50. Ibid., 31.

51. Ibid., 53.

52. For a history of the city-body symbolic form, see Sennett, *Flesh and Stone*.

53. Grosz, "Bodies-Cities," 47.

54. Merriam-Webster Medical Desk Dictionary, s.v. "anhedonia," accessed 22 October 2011, http://www2.merriam-webster.com/cgi-bin/mwmedsamp.

55. Wallace, *Brief Interviews with Hideous Men*, 49.

56. Jameson, *Postmodernism*, 53.

57. Smith, *Uneven Development*, 45.

58. Soja, *Postmodern Geographies*, 57.

59. Grover, *North Enough*, 18.

60. Grosz, "Bodies-Cities," 43.
61. Ibid.
62. Jameson, *Postmodernism*, 38.
63. Ibid., 39.
64. Stephen J. Burn examines the "bodily gestalts [that] are a recurring feature of Wallace's fiction." Burn, "'Webs of Nerves Pulsing and Firing,'" 64. Burn uses this observation to launch a discussion of the theories of mind that circulate in Wallace's writings.
65. The Student Union enters the narrative because it houses WYYY, the studio where Joelle van Dyne records her radio program, the "Madame Psychosis Hour."
66. Kolodny, *The Lay of the Land*; and Westling, *The Green Breast of the New World*.
67. Soja, *Postmodern Geographies*, 133.
68. Forster, *Howards End*, 105.
69. Buell, *Writing for an Endangered World*, 89.
70. Simmel, "The Metropolis and Mental Life," 12.
71. Ibid., 15.
72. Ibid.
73. Buell, *Writing for an Endangered World*, 89.
74. Ibid.
75. Wallace, "Interview with David Foster Wallace," 127.
76. Russell, "Some Assembly Required," 157.
77. Ibid., 166.
78. I assign this particular thought to Mario, but note that focalization shifts away from this character with the use of "bullshit" and the third-person reference to him ("boy").
79. Buell, *Writing for an Endangered World*, 108.
80. de Certeau, *The Practice of Everyday Life*, xxiv.
81. Russell, "Some Assembly Required," 164.
82. Ibid., 166.
83. Simmel, "The Metropolis and Mental Life," 15.
84. Palumbo-Liu, *The Deliverance of Others*, 13–15. One of Palumbo-Liu's departures for thinking through affective invasion is Teresa Brennan's eponymous study of the "transmission of affect." Brennan identifies how the communicability of emotion draws attention to our vulnerability to our human surroundings.
85. Simmel, "The Metropolis and Mental Life," 15.
86. Pole, "Disgust and Other Forms of Aversion," 221.

87. Kakutani, "A Country Dying of Laughter."
88. This point hints at D. A. Miller's question of whether style, because it is so conspicuous—even flamboyant, is excessive by nature. See Miller, *Jane Austen*, 8, 18.
89. Quoted in Max, *Every Love Story Is a Ghost Story*, 281.
90. Kristeva, *Powers of Horror*, 4.
91. Ibid., 12.
92. Ibid., 53.
93. Rozin, Haidt, and McCauley, "Disgust," 757.
94. Wallace's exhaustively titled short story, "On His Deathbed, Holding Your Hand, the Acclaimed New Young Off-Broadway Playwright's Father Begs a Boon," confirms this point. As the narrator deteriorates from a cluster of diseases, he confesses that his newborn son's too-bodily body disgusted him. The child teaches the father "to despise the body, what it is to have a body—to be disgusted, repulsed." Wallace, *Brief Interviews with Hideous Men*, 259. "What it is to have a body" was a source of trouble for the anxiety-prone Wallace. He initially donned his signature white bandana to control his profuse sweating and sometimes took multiple showers a day. Avid readers of Wallace know that a lot of sweat dampens the pages of *Infinite Jest* and his stories.
95. Quoted in Ahmed, *The Cultural Politics of Emotion*, 83.
96. Powers, *The Echo Maker*, 166.
97. Catherine Nichols argues that carnivalesque elements such as irony, metafiction, and polyphonic intertextuality serve "not only to de-center empty avant-gardism . . . but to defamiliarize the hallmarks of classic realism." Nichols, "Dialogizing Postmodern Carnival," 14. She concurs with Tom LeClair, who holds that narrative techniques such as "multiple points of view, both first- and third-person; stylistic tours de force in several dialects; a swirling associative structure; and alternations in synecdochic scale" distance the novel from the "'soothing, familiar and anesthetic'" packaging of realism. LeClair, "The Prodigious Fiction of Richard Powers," 35.
98. Quoted in Max, *Every Love Story Is a Ghost Story*, 173.
99. Probyn, "Writing Shame," 86.
100. Focusing on disgust also takes us outside the novel's own seductive mind. *Infinite Jest* is, like the film it imagines, "so bloody compelling" that the scholarship on it can get stuck in the text's own terms of recursivity, solipsism, irony, and undecidability (839).
101. Miller's study, published the year after *Infinite Jest*, appears in the David Foster Wallace Library at the Harry Ransom Center at the University

of Texas at Austin. Wallace's underlining, asterisks, and other annotations show that his fascination with the emotion continued as he began work on future stories and *The Pale King* (2011). Several themes especially attract Wallace's pen: the idea of disgust as an assertion of the subject's superiority "that at the same time recognizes the vulnerability of that superiority to the defiling powers of the low"; the bodily orifices that produce disgust (anus, vagina, nose, etc.); and the claim that "to feel disgust is human and humanizing." Miller, *The Anatomy of Disgust,* 9, 11.

102. Ibid., 180–81.

103. Of course, this universality is illusory. We need only consider that food tolerances differ so radically across cultures: one man's delicacy is another's emetic. Still, we expect disgust to "work" within given cultural parameters and trust that our body's immediate responses may model a norm for all.

104. Ngai, *Ugly Feelings,* 336.

105. Ibid., 354.

106. Cousins, "The Ugly (Part 1)," 64.

107. Kristeva, *Powers of Horror,* 8.

108. Hobbes, *Leviathan,* 38.

109. Miller, *The Anatomy of Disgust,* x.

110. Bourdieu, *Distinction,* 494.

111. Ahmed, *The Cultural Politics of Emotion,* 85.

112. Pole, "Disgust and Other Forms of Aversion," 225. In this respect, disgust shares the general oscillatory form of wonder as analyzed in chapter 3.

113. Ibid.

114. Ibid.

115. Kristeva, *Powers of Horror,* 3, 10.

116. Ahmed, *The Cultural Politics of Emotion,* 89.

117. Ricoeur, *Time and Narrative,* 79.

118. The phrase appears in Hal's seventh-grade essay on the evolving role of the TV hero. He predicts that the individual who will supersede the "'post'-modern" hero will be "a hero of non-action, the catatonic hero, the one beyond calm, divorced from all stimulus" (142).

119. Wallace, *This Is Water,* 120.

120. Thrailkill, *Affecting Fictions,* 51.

121. On this point, the play between humor and disgust in *Infinite Jest* requires comment. I believe that the humor of the novel dissipates over time under the pressure of its content. As Wallace told David Lipsky right after the novel came out, it's "supposed to be sort of fun and unfun. For

instance, I like a joke that you laugh hard at, but then it's sort of unsettling, and you think about it for a while. It's not quite black humor, but it's a kind of, a kind of creepy humor." Lipsky, *Although Of Course You End Up Becoming Yourself*, 272. Humor no longer dominates as scenes of pain and anguish accumulate. Flooded by passages depicting pain, engaged readers reflect on why stuff that had once been so funny now feels sometimes disgusting, sometimes sad.

122. Thrailkill, *Affecting Fictions*, 51.

5. THE ANXIETY OF INTERVENTION IN LESLIE MARMON SILKO AND MARGE PIERCY

1. The early 2000s see a return of anxieties about nuclear weapons in the U.S. that cooled down with the gradual dismantling of the communist bloc in the late 1980s and early 1990s. These anxieties reemerge and shift to the new centers of U.S. geopolitical and military engagement: the Middle and Far East.

2. Andrew Ross examines how urban renewal projects in New York City stigmatized and displaced the poor and the sick in the 1960s and 1970s. "In public language," he argues, "'urban problems' has become a codeword for race." Ross, *The Chicago Gangster Theory of Life*, 138.

In an essay Piercy refers to cities as "grids of Them-and-Us experiences" built around cultural and economic distinctions. Piercy, "The City as Battleground," 209. She relates a childhood memory of the 1943 Detroit race riot, which set a precedent for the wave of race-motivated uprisings in the 1960s and 1970s that occurred while Piercy was writing *Woman on the Edge of Time*.

3. Piercy, *Woman on the Edge of Time*, 190. Hereafter cited parenthetically. Mattapoisett is a Wampanoag word for "resting place" and is a town on the Cape in Massachusetts.

4. Though the chart puts the main diegesis under suspicion, it is inaccurate; it misstates Connie's heritage, designating her Puerto Rican rather than Mexican. For this reason and because it comes from the reviled doctors, it does not have the final word on Connie's condition or the preceding narrative.

5. For Jameson, the "fundamental anxiety of Utopia" is "the fear of losing that familiar world in which all our vices and virtues are rooted (very much including the very longing for Utopia itself) in exchange for a world in which all these things and experiences—positive as well as negative—will have been obliterated." Jameson, *Archaeologies of the Future*, 97.

6. Anderson, "The River of Time," 76.

7. Stability does not mean universal harmony and happiness. Interpersonal and internecine conflicts do exist in Mattapoisett, most notably a battle between Luciente's people and the "few but determined" members of a rapacious society of "androids, robots, [and] cybernauts" (261).

8. For a primer on neoliberal economics, see Harvey, *A Brief History of Neoliberalism*.

9. Silko, *Almanac of the Dead*, 569, 475. Hereafter cited parenthetically.

10. Huhndorf, *Mapping the Americas*, 141.

11. Joni Adamson, "'¡Todos Somos Indios!'" 6. T. V. Reed shares Adamson's interest in the compatibility of the environmental justice and decolonization movements in *Almanac*. Reed, "Toxic Colonialism." On the novel's environmental politics, see also Ammons, *Brave New Words*; Barber, "Wisecracking Glen Canyon Dam," 127–43; Bowers, "Eco-Criticism in a (Post-)Colonial Context"; O'Meara, "The Ecological Politics"; and Teale, "The Silko Road from Chiapas."

12. Several characters iterate Old Yoeme's message. Calabazas, a wizened Mexican-Yaqui drug smuggler; Clinton, "the first Black Indian" and a leader of the Army of the Homeless (404); Tacho and his twin, El Feo; and the preachers at the International Holistic Healers Convention.

13. Silko, "The Fourth World," 125.

14. Heise, *Sense of Place and Sense of Planet*, 29.

15. Ibid., 33.

16. Chapter 2 identifies the residue of this desire in David Wojnarowicz's quasi-Emersonian yearnings for fluidity between self and other beings.

17. Allen, *The Sacred Hoop*, 119.

18. Silko, *Ceremony*, 116.

19. Reed delineates three forms of restoration in *Almanac*: the return of the lands to tribes, the feeling of sacred connection to and stewardship for those lands, and the vitality of the Earth writ large. Reed, "Toxic Colonialism."

20. *Almanac* has a pantribal reach. All Indian people—from the Yaquis in the southwest U.S. to the Eskimos in the Arctic Circle—have tribally specific stories and relations to the land, but they share affinities with others. For readings of the novel's pantribalism and transamericanism, see Adamson, *American Indian Literature*; Huhndorf, *Mapping the Americas*; Jarman, "Exploring the World of the Different"; Karem, *The Romance of Authenticity*; Krupat, *Turn to the Native*; Olmsted, "The Uses of Blood"; and Sadowski-Smith, *Border Fictions*.

21. The idea of a naturalized connection between people and land evokes N. Scott Momaday's controversial trope of "blood memory." According to Chadwick Allen, this trope involves "the assertion of an unmediated relationship to indigenous land bases . . . the continuation of oral traditions . . . and the power of the indigenous writer's imagination to establish communion with ancestors." Allen, *Blood Narrative*, 178. Allen notes that Arnold Krupat and others have objected that the blood memory trope is "'absurdly racist.'" Ibid., 271n45. *Almanac* potentially circumvents this criticism due to its pantribalism.

22. *Almanac* is made up of six parts, which contain between one and eight books that are further divided into chapters.

23. As in *Ceremony*, the estranged male character in *Almanac* reestablishes his connection to place and tribe when he recognizes the role of uranium mining in tribal and global history. Tayo trespasses onto a government-owned mine, and his link to both the Laguna people and Japanese victims of the atomic bomb overwhelms him. This epiphany completes his healing ceremony. *Almanac*, by comparison, depicts mining after the worldwide collapse in uranium prices in the 1980s that forced the temporary closure of all U.S. open-pit mines in 1992. The mines were left, serving as reminders of environmental injustice.

24. I return to *Almanac*'s ending and question the closure it provides in the final section of this chapter.

25. For a critique of *Almanac*'s leveling of "the complexity of [Europeans'] moral and spiritual history," see Garrard, "Ecocriticism," 55.

26. Indigenous Alliance of the Americas on 500 Years of Resistance, "Declaration of Quito."

27. See, for example, the parodic list of events on the convention schedule (717). *Almanac* critics tend to take the event at face value, but the episode's farcical aspects undermine any purely earnest reading.

28. "No Compromise in Defense of Mother Earth."

29. Silko relates the origins of Lecha's character in "Notes on *Almanac of the Dead*." Silko, *Yellow Woman*, 137–38. This essay mentions other correspondences between people and events in the novel and in Silko's own life.

30. Bowers, "Eco-Criticism in a (Post-)Colonial Context," 275–76.

31. Rose, *The Politics of Life Itself*, 83.

32. Ibid., 3.

33. See Mogen for a reading of *Almanac* as a work of "scientific speculation." Mogen, "Native American Visions of Apocalypse," 159.

34. Adamson, *American Indian Literature*, 169.

35. Silko relocated to the Tucson area from Laguna, New Mexico in 1978 and, with support from a MacArthur Foundation "Genius" Grant, wrote *Almanac* there.

36. Allen, *Biosphere 2*, 10.

37. Suplee, "Brave Small World."

38. For official histories of the project, see ibid.; Allen, *Biosphere 2*; and Alling, Nelson, and Silverstone, *Life Under Glass*. The following newspaper accounts detail and critique the project: Chandler, "An Earthbound Ark Sets Sail"; Lew, "Peering into a World Under Glass"; Toufexis, "The Wizards of Hokum"; and Walker, "Home, Sweet Biome."

39. For example, Chandler, "An Earthbound Ark Sets Sail."

40. Cooper, "Profits of Doom," 31–32.

41. On *Almanac*'s depiction of coalition building, see Adamson, "'¡Todos Somos Indios!'"; Olmsted, "Uses of Blood"; O'Meara, "The Ecological Politics"; Reed, "Toxic Colonialism"; Romero, "Envisioning a 'Network of Tribal Coalitions'"; and Sadowski-Smith, *Border Fictions*.

42. Admittedly, both Clinton and environmentalism come across as caricatures here. Later sections of this chapter elaborate on characterization in *Almanac*.

43. Clinton's views resonate with social theorist Andrew Ross's analysis of anti-immigration groups in the southwest U.S. who are "tapping a deep vein of public anxiety that connects the defense of pristine resources to the defense of racial purity." Ross, *Bird on Fire*, 192. See also *The Chicago Gangster Theory of Life*, 237–73.

44. Devall and Sessions, *Deep Ecology*, 70.

On the population question, *Almanac* again refracts contemporary debates. Between the 1970s and 1990s, the overpopulation issue fractured major environmental groups, the Sierra Club most notably, because population control was too easily associated with racism. Jennifer Ludden substantiates this fear in her report on how anti-immigration groups couch their stances in environmentalist rhetoric. Ludden, "Ads Warn That Immigration Must Be Reduced." For accounts of the schism within the Sierra Club, see Knickerbocker, "A 'Hostile' Takeover Bid at the Sierra Club"; and Davila, "Immigration Dispute Spawns Factions." Ian Angus and Simon Butler have recently surveyed and critiqued the "myth of overpopulation." Angus and Butler, *Too Many People?* ix.

45. Angus and Butler, *Too Many People?* 115.

46. Guha, "Radical American Environmentalism," 74.

47. Ibid., 75, 74n2.

48. Waldby, *The Visible Human Project*, 136.

49. Ibid.

50. Rebecca Tillett develops this line of argument through a reading of Beaufry, whose "assurance of his own superiority both parodies and exposes the ideological biases within the policies by which science, technology, and industry organize their hierarchies of social and natural significance." Tillett, "Reality Consumed by Realty," 157–58.

51. For a reading of how disability and homosexuality constitute borderland identities, see Jarman, "Exploring the World of the Different." I share Jarman's interest in *Almanac*'s eugenic tropes, but disagree that those with "alternative" identities necessarily offer efficacious models of agency in the novel. Ibid., 161.

52. Stein, *Shifting the Ground*, 210.

53. Ibid.

54. Stanford, "'Human Debris,'" 26.

55. Waldby, *The Visible Human Project*, 18.

56. Rabinow and Rose, "Biopower Today," 198.

57. *Almanac*'s critique of technology is not absolute. Digital and communications technologies serve the revolutionary projects of Zeta, Lecha, and others. Televisions are conduits for Native prophecies, and computers preserve the ancient almanacs. Most importantly, Zeta enlists South Korean programmer, Awa Gee, in the transamerican revolution. Gee develops algorithms to spy on the enemies, hacks into the electrical grid, and builds detonators for the bombs Ferro sets off throughout Tucson. While Zeta has misgivings about the latter project (736), the novel endorses technological interventions as tools for resistance so long as they do not alter the very matter of life.

58. Tillett, "Reality Consumed by Realty," 156.

59. Hacking, "Our Neo-Cartesian Bodies in Parts," 78.

60. Ibid., 79.

61. Ibid., 80.

62. Heise, "The Hitchhiker's Guide to Ecocriticism," 509.

63. Barkun, "Divided Apocalypse," 263.

64. Ibid.

65. Ibid., 274.

66. For studies of apocalypticism in American thought and culture, see Boyer, *When Time Shall Be No More*; Buell, *From Apocalypse to Way of Life*; Dewey, *In a Dark Time*; Robinson, *American Apocalypses*; and Zamora, *Writing the Apocalypse*. Apocalypticism's role in Anglo-American environmental

discourse is addressed in Buell, *The Environmental Imagination*; and Garrard, *Ecocriticism*.

67. A handful of times, *Almanac* references the Ghost Dance religion, an apocalyptic American Indian belief system that emerged in the U.S. West in the 1880s. At the Healers Convention, the Lakota Wilson Weasel Tail reminds the audience of the Ghost Dance that believers performed to connect them with ancestral spirits. According to the Lakota strain of the religion, the ancestors would converge into a "spirit army" that would overthrow white rule and restore tribal lands to their rightful stewards (724). Given the novel's apocalypticism, it is perhaps surprising that the Ghost Dance does not figure as prominently here as it does in Silko's next novel, *Gardens in the Dunes* (1999). However, because the Ghost Dance syncretized Christian and Native spirituality, it might not have fit with *Almanac*'s censure of Western belief systems.

68. See Revelations 6:12, 16:2–4, and 16:12–18 for references to these same catastrophes.

69. On this point, we can think back to the serial killer anecdote analyzed previously, in which "tending" becomes a euphemism for murder.

70. Silko, *Yellow Woman*, 139.

71. As Patricia Limerick notes, development in the "new West" often grows out of the idea that its so-called open spaces can remedy physiological, societal, and psychological ailments: "dry air to cure respiratory problems, open spaces to relieve the pressures that weigh down the soul in teeming cities, freedom and independence to provide a restorative alternative to mass society's regimentation and standardization." Limerick, *Something in the Soil*, 277. These hopes presuppose the taming of the West, in particular its forbidding deserts and mountains, that Leah Blue undertakes. On patterns of greenwashed politics and real estate development in early twenty-first-century Arizona, see Ross, *Bird on Fire*.

72. To support her claims that "people wanted to have water around them in the desert," Leah earlier cites "market research [that] had repeatedly found new arrivals in the desert were reassured by the splash of water" (374–75). The fictional Venice, Arizona evokes a turn-of-the-century development in Southern California also named Venice, a "lagoon-laced Los Angeles suburb keyed to the Mediterranean metaphor." Starr, *California*, 151.

73. Quoted in Tuan, *Topophilia*, 67. For rejoinders to desert "place bashing," see Glotfelty, "Literary Place Bashing"; and Lynch, *Xerophilia*.

74. On this representational tendency, see Sontag, *Illness as Metaphor and AIDS and Its Metaphors*, 73.

75. Harris Feinsod offered this helpful gloss.

76. While it is true that there is almost no positive expression of sexuality in *Almanac*, the novel's impenitent use of male same-sex desire to mark moral turpitude (in Serlo and Beaufry, David, and others) undermines the novel's critique of those who do the same with racial, ethnic, and class positioning.

77. Massumi, "Preface," viii.

78. Several scholars address *Almanac*'s engagement with Mayan texts and prophecy. See Adamson, *American Indian Literature*; Fischer-Hornung, "Economics of Memory"; Horvitz, "Freud, Marx, and Chiapas"; Mogen, "Native American Visions of the Apocalypse"; Olmsted, "Uses of Blood"; Reinecke, "Overturning the (New World) Order"; and Van Dyke, "From Big Green Fly to the Stone Serpent."

79. Killingsworth and Palmer, "Millennial Ecology," 22.

80. Ibid., 23.

81. Ibid., 41.

82. Buell, *The Environmental Imagination*, 285.

83. Bercovitch, *The American Jeremiad*, xi.

84. Quoted in ibid., 5.

85. Ibid., 20.

86. My thanks to Kiara Vigil for pointing me to Apess's writing.

87. The last section of this chapter addresses how Silko's novel also innovates on its titular genre, Mayan and early American almanacs.

88. Bercovitch, *The American Jeremiad*, 23.

89. Ibid., 191.

90. Buell, *The Environmental Imagination*, 308.

91. Several critics, most notably Adamson, have elaborated on *Almanac*'s "quality of prophecy." Adamson, *American Indian Literature*, 128. In particular, they note the correspondence between the novel and the Zapatista liberation group that emerged in Chiapas, Mexico in 1994 to combat the decimation of Mayan lands and culture and to influence Mexican politics. See also Adamson, "'¡Todos Somos Indios!'"; Horvitz, "Freud, Marx, and Chiapas"; Romero, "Envisioning a 'Network of Tribal Coalitions'"; and Sadowski-Smith, *Border Fictions*. In "An Expression of Profound Gratitude to the Zapatistas, January 1, 1994," Silko situates the Zapatista rebellion as a continuation of her ancestors' stories of resistance to the "destroyers." Silko, *Yellow Woman*, 153.

92. Bercovitch, *The American Jeremiad*, 23.

93. Ngai, *Ugly Feelings*, 29.

94. Bloch, *The Principle of Hope*, 75.

95. Ibid., 74.
96. Ibid.
97. Ibid., 108.
98. Ngai, *Ugly Feelings*, 215.
99. Ibid., 236, 246.
100. Ibid., 212.
101. See Kierkegaard, *The Concept of Anxiety*, 42; and Ngai, *Ugly Feelings*, 232, on Heidegger. Ngai elaborates a genealogy of the claim that anxiety "has no concrete thing or 'entity-within-the-world' as its object." Ibid., 393n35.
102. Bloch, *The Principle of Hope*, 3.
103. Killingsworth and Palmer, "Millennial Ecology," 23.
104. Ngai, *Ugly Feelings*, 393n35.
105. St. John, "Almanac of the Dead," 124.
106. Jones, "Reports from the Heartland," 81.
107. Knickerbocker, "Dark Beauty, Bright Terror," 13.
108. As Adamson has shown, the novel also makes readers uncomfortable because it violates their expectations for an "authentic" Native American novel. Adamson, *American Indian Literature*, 130. Depicting "genocide, resistance, and rebellion," *Almanac* speaks to ethnically and racially coded discomfort with Native activists' place in the literary canon and in politics. Ibid., 134.
109. One citizen reviewer on Amazon.com describes it as "jumpy," which could refer either to the plot or to its effect on the reader. Kuzyk, "An Excellent Book . . ."
110. A text can, of course, misfire even more drastically and miss its target. Even in this case the critic's dismissal of *Almanac* depends on the enormity of the book's intentions. Sven Birkerts roundly and controversially dismisses Silko's vision: "That the oppressed of the world should break their chains and retake what's theirs is not an unappealing idea (for some), but it is so contrary to what we know both of the structures of power and the psychology of the oppressed that the imagination simply balks. . . . [Silko's] premise of revolutionary insurrection is tethered to airy nothing. It is, frankly, naive to the point of silliness." Birkerts, "Apocalypse Now," 41.
111. Tallent, "Storytelling with a Vengeance."
112. Buell, *From Apocalypse to Way of Life*, 201.
113. Ngai, *Ugly Feelings*, 246.
114. Rose, *The Politics of Life Itself*, 51.
115. Ibid.
116. Ibid.
117. Bloch, *The Principle of Hope*, 75.

118. Buell, *From Apocalypse to Way of Life*, 201; Zamora, *Writing the Apocalypse*, 76.

119. Reed, "Toxic Colonialism," 37.

120. Silko, "An Interview with Leslie Marmon Silko (1994)," 145.

121. Ibid., 144.

122. Ibid. Brackets in original.

123. It's worth noting that, in a 1992 interview, Silko retracts the idea of radical revolution entirely, declaring, "'the only revolution I truly believe in is one of awareness of perception. At the end of my novel, I try to show the awareness or the change moving to be something organic, a part of this continent and something inclusive, which is the American Indian way. I would hope it would be a gentle revolution over hundreds and hundreds of years.'" Kelleher, "Predicting a Revolt to Reclaim the Americas."

124. For another example, see one paragraph on page 264: "Menardo sits back . . . Menardo checks his reflection . . . Menardo sees a man . . . Menardo refuses to be stared down . . . Menardo drops his eyes . . . Menardo admires the parachute."

125. Michaels, *The Shape of the Signifier*, 24.

126. Ibid., 149.

127. Ibid., 11.

128. Ibid., 135. For alternate readings of Bartolomeo's execution, see Adamson, *American Indian Literature*, 156; and Huhndorf, *Mapping the Americas*, 160–61.

129. Michaels, *The Shape of the Signifier*, 23.

130. *OED Online*, s.v. "almanac, n.," Oxford University Press, last modified September 2012, http://www.oed.com/view/Entry/5564?rskey=ghfTeI&result=1&isAdvanced=false. *Almanac of the Dead* offers the following definition for its eponym:

1. almanakh: Arabic.
2. almanac: A.D. 1267 English from the Arabic.
3. almanaque: A.D. 1505 Spanish from the Arabic.
4. a book of tables containing a calendar of months and days with astronomical data and calculations.
5. predicts or foretells the auspicious days, the ecclesiastical and other anniversaries.
6. short glyphic passages give the luck of the day.
7. Madrid
   Paris   Codices
   Dresden   (136)

131. My point accords with Adamson's reading of the almanac as a "challenge [to] the very notion of authoritative discourse." Adamson, *American Indian Literature*, 134.

132. Critics who read the novel as restorative tend to focus their analysis on this last chapter. See, for example, Norden, "Ecological Restorations."

133. Ngai, *Ugly Feelings*, 246–47.

134. Ibid., 245.

CONCLUSION: HOW DOES IT FEEL?

1. Ward, "NOAA's 2012 Arctic 'Report Card.'"

2. See Adamson, "For the Sake of the Land and All People"; and Alaimo, "The Trouble with Texts."

3. Hawken, *Blessed Unrest*, 7.

4. Ibid., 8. Other recent treatises on environmental hope include Alperovitz, *America Beyond Capitalism*; Solnit, *Hope in the Dark*; Speth, *America the Possible*; and Wohlforth, *The Fate of Nature*.

5. Wallace, *Infinite Jest*, 694.

6. Sharot, Korn, and Dolan, "How Unrealistic Optimism Is Maintained," 1475.

7. Mims, "80 Percent of Humans Are Delusionally Optimistic."

8. Ehrenreich, *Bright-Sided*, 17.

9. Ibid., 78.

10. Quoted in ibid. Ehrenreich goes on to point out the residues of Calvinism in the movements that took shape around positive thinking in the nineteenth century. Ibid., 69.

11. Powers, "What Does Fiction Know?"

12. Ehrenreich, *Bright-Sided*, 8. Lauren Berlant's concept of "cruel optimism" similarly reassesses optimism but from the perspective of cultural affect studies. As I noted in chapter 1, cruel optimism is a form of positivity that arises from attachments to objects, ideologies, and behaviors that are ultimately harmful, and it leads to forms of "slow death" for the individual and the populations to which she belongs. Berlant, *Cruel Optimism*, 102.

13. Cvetkovich, *Depression*, 15.

14. Ibid., 2.

15. Muñoz, *Cruising Utopia*, 12. I cluster Cvetkovich and Muñoz together here, but the former places more emphasis on "ordinary form[s] of spiritual practice that I call *the utopia of everyday habit*," habits such as bodywork, exercise, craft making, personal writing, and even flossing. Cvetkovich,

Depression, 159. While some of these practices are public (for example, producing crafts in and for gallery spaces and fairs), many primarily redound to the individual and are part of an alchemy of "self-transformation." Ibid., 168. Muñoz's optimism of the negative, on the other hand, enlists queer artworks and political acts to speculate on the possibility for collective utopianism.

16. Hacking, "Our Neo-Cartesian Bodies in Parts," 80.
17. Speth, *America the Possible*, 194.
18. Powers, "What Does Fiction Know?"
19. There has been a "turn to the planet" within globalization and cosmopolitan studies as well as environmental studies. A genealogy of the former strands of cultural criticism includes the work of Masao Miyoshi, Wai Chee Dimock, and Gayatri Spivak, among others. In 2001 Masao plotted this turn in an essay that announced that "literature and literary studies now have one basis and goal: to nurture our common bonds to the planet—to replace the imaginaries of exclusionist familialism, communitarianism, nationhood, ethnic culture, regionalism, 'globalization,' or even humanism, with the ideal of planetarianism." Miyoshi, "Turn to the Planet," 295. See also Dimock and Buell, *Shades of the Planet*; and Spivak, *Death of a Discipline*.
20. For reflections on these debates, see the first two chapters of Buell, *The Future of Environmental Criticism*; Oppermann, "Ecocriticism's Theoretical Discontents"; and Phillips, *The Truth of Ecology*. Simon C. Estok and S. K. Robisch revived this debate about whether theory amounts to obscurantism in the pages of *ISLE* in 2009. Estok, "Theorizing in a Space of Ambivalent Openness"; and Robisch, "The Woodshed." In his rejoinder to theory aversion, Estok urges ecocritics to privilege "ecophobia" as an analytic. I agree with Estok that a major task of ecocriticism lies in "identifying the affective ethics a text produces . . . [and] having the willingness to listen to, to think about, and to see the values that are written into that work through the representations of nature we imagine, theorize, and produce." Estok, "Reading Ecophobia," 76. However, I do not agree that we should privilege fear over all emotions.
21. 350.org targets the fossil fuel economy in order to solve the climate crisis while Speth addresses a wider range of social, economic, and environmental transformations required to usher in a more just political economy. See the list in Speth, *America the Possible*, 10–11.
22. Critical Art Ensemble, "GenTerra."
23. Critical Art Ensemble, "Fear and Profit in the Fourth Domain."
24. Ibid.
25. The Cultural Cognition Project, "Home."

26. Kahan et al., "The Second National Risk and Culture Study."
27. Ibid., 1.
28. Ibid., 5.
29. Altieri, *The Particulars of Rapture*, 254.
30. Wallace, *Girl with Curious Hair*, 333.

WORKS CITED

Abbey, Edward. *Desert Solitaire: A Season in the Wilderness*. New York: Touchstone, 1990 [1968].
"About Adbusters." *Adbusters.org*. Adbusters Media Foundation. Accessed 27 August 2011. http://www.adbusters.org/about/adbusters.
"About Sandhill Cranes." *OutdoorNebraska.ne.gov*. Nebraska Game and Parks Commission. Accessed 21 September 2011. http://outdoornebraska.ne.gov/conservation/wildlife-viewing/SandhillCranes/sandhill.asp.
Adamson, Joni. *American Indian Literature, Environmental Justice, and Ecocriticism: The Middle Place*. Tucson: University of Arizona Press, 2001.
———. "For the Sake of the Land and All People: Teaching American Indian Literatures from an Environmental Justice Perspective." In *Teaching North American Environmental Literature*, ed. Laird Christensen, Mark C. Long, and Fred Waage, 194–202. New York: Modern Language Association, 2008.
———. "'¡Todos Somos Indios!': Revolutionary Imagination, Alternative Modernity, and Transnational Organizing in the Work of Silko, Tamez, and Anzaldúa." *Journal of Transnational American Studies* 4, no. 1 (2012): 1–26.
Adamson, Joni, Mei Mei Evans, and Rachel Stein, eds. *The Environmental Justice Reader: Politics, Poetics, and Pedagogy*. Tucson: University of Arizona Press, 2002.
Adorno, Theodor. *Negative Dialectics*. New York: Routledge, 1990 [1973].
Ahmed, Sara. *The Cultural Politics of Emotion*. New York: Routledge, 2004.

Alaimo, Stacy. *Bodily Natures: Science, Environment, and the Material Self.* Bloomington: Indiana University Press, 2010.
———. "The Trouble with Texts; or, Green Cultural Studies in Texas." In *Teaching North American Environmental Literature*, ed. Laird Christensen, Mark C. Long, and Fred Waage, 369–76. New York: Modern Language Association, 2008.
———. *Undomesticated Ground: Recasting Nature as Feminist Space.* Ithaca, NY: Cornell University Press, 2000.
"All About BALANCE." *Balance.org.* Population-Environment Balance. 2001. Accessed 1 April 2012. http://www.balance.org/about.html.
Allen, Chadwick. *Blood Narrative: Indigenous Identity in American Indian and Maori Literary and Activist Texts.* Durham, NC: Duke University Press, 2002.
Allen, John. *Biosphere 2: The Human Experiment*, ed. Anthony Blake. New York: Penguin, 1991.
Allen, Paula Gunn. *The Sacred Hoop: Recovering the Feminine in American Indian Traditions.* Boston: Beacon, 1986.
Alling, Abigail, Mark Nelson, and Sally Silverstone. *Life Under Glass: The Inside Story of Biosphere 2.* Oracle, AZ: Biosphere, 1993.
Alperovitz, Gar. *America Beyond Capitalism: Reclaiming Our Wealth, Our Liberty, and Our Democracy.* Hoboken, NJ: Wiley, 2005.
Alpers, Paul. *What Is Pastoral?* Chicago: University of Chicago Press, 1996.
Altieri, Charles. *The Particulars of Rapture: An Aesthetics of the Affects.* Ithaca, NY: Cornell University Press, 2003.
Ammons, Elizabeth. *Brave New Words: How Literature Will Save the Planet.* Iowa City: University of Iowa Press, 2010.
Anderson, Amanda. *The Way We Argue Now: A Study in the Cultures of Theory.* Princeton: Princeton University Press, 2005.
Anderson, Eric Gary. *American Indian Literature and the Southwest: Contexts and Dispositions.* Austin: University of Texas Press, 1999.
Anderson, Perry. "The River of Time." *New Left Review* 26 (March/April 2004): 67–77.
Angus, Ian, and Simon Butler. *Too Many People? Population, Immigration, and the Environmental Crisis.* Chicago: Haymarket, 2011.
Antonetta, Susanne. *Body Toxic: An Environmental Memoir.* Washington, DC: Counterpoint, 2001.
Atwood, Margaret. "In the Heart of the Heartland." Review of *The Echo Maker*, by Richard Powers. *New York Review of Books*, 21 December 2006.

Augé, Marc. *Non-Places: Introduction to an Anthropology of Supermodernity*. 1992. Trans. John Howe. New York: Verso, 1995.

Barad, Karen. *Meeting the Universe Halfway: Quantum Physics and the Entanglement of Matter and Meaning*. Durham, NC: Duke University Press, 2007.

Barber, Katrine E. "Wisecracking Glen Canyon Dam: Revisioning Environmentalist Mythology." In *Change in the American West: Exploring the Human Dimension*, ed. Stephen Tchudi, 127–43. Reno: University of Nevada Press, 1996.

Barbour, Michael G. "Ecological Fragmentation in the Fifties." In *Uncommon Ground: Rethinking the Human Place in Nature*, ed. William Cronon, 233–55. New York: Norton, 1995.

Barkun, Michael. "Divided Apocalypse: Thinking About the End in Contemporary America." *Soundings* 65, no. 3 (1983): 257–80.

Bate, Jonathan. *Romantic Ecology: Wordsworth and the Environmental Tradition*. New York: Routledge, 1991.

Bauman, Zygmunt. *Postmodern Ethics*. Malden, MA: Blackwell, 1993.

"A Beautiful History." *Kab.org*. Keep American Beautiful, Inc. 2006. Accessed 21 October 2011. http://www.kab.org/site/PageServer?pagename=kab_history.

Beck, Ulrich. *Risk Society: Towards A New Modernity*. 1986. Trans. Mark Ritter. London: Sage, 1992.

Bell, David, and Gill Valentine. "Queer Country: Rural Lesbian and Gay Lives." *Journal of Rural Studies* 11, no. 2 (1993): 113–22.

Bell, Virginia E. "Counter-Chronicling and Alternative Mapping in *Memoria del fuego* and *Almanac of the Dead*." *MELUS* 25, no. 3–4 (2000): 5–30.

Bennett, Jane. *Vibrant Matter: A Political Ecology of Things*. Durham, NC: Duke University Press, 2010.

Bercovitch, Sacvan. *The American Jeremiad*. Madison: University of Wisconsin Press, 1978.

Berlant, Lauren. *Cruel Optimism*. Durham, NC: Duke University Press, 2011.

Berleant, Arnold. *The Aesthetics of Environment*. Philadelphia: Temple University Press, 1992.

Best, Stephen, and Sharon Marcus. "Surface Reading: An Introduction." *Representations* 108, no. 1 (2009): 1–21.

Biehl, João. *Will to Live: AIDS Therapies and the Politics of Survival*. Princeton: Princeton University Press, 2007.

Birkerts, Sven. "Apocalypse Now." Review of *Almanac of the Dead*, by Leslie Marmon Silko. *New Republic*, 14 November 1991, 39–41.

Bloch, Ernst. *The Principle of Hope*, vol. 1. Trans. Neville Plaice, Steven Plaice, and Paul Knight. Cambridge: MIT Press, 1995.

Botkin, Daniel B. *Discordant Harmonies: A New Ecology for the Twenty-First Century*. New York: Oxford University Press, 1990.

Bourdieu, Pierre. *Distinction: A Social Critique of the Judgment of Taste*. Trans. Richard Nice. Cambridge: Harvard University Press, 1984.

Bowers, Maggie Ann. "Eco-Criticism in a (Post-)Colonial Context and Leslie Marmon Silko's *Almanac of the Dead*." In *Towards a Transcultural Future: Literature and Human Rights in a "Post"-Colonial World*, ed. Peter H. Marsden and Geoffrey V. Davis, 267–76. Amsterdam: Rodopi, 2004.

Boyd, Brian. *On the Origin of Stories: Evolution, Cognition, and Fiction*. Cambridge: Belknap Press of Harvard University Press, 2009.

Boyer, Paul. *When Time Shall Be No More: Prophecy Belief in Modern American Culture*. Cambridge: Belknap Press of Harvard University Press, 1992.

Brennan, Teresa. *The Transmission of Affect*. Ithaca, NY: Cornell University Press, 2004.

Brigham, Ann. "Productions of Geographic Scale and Capitalist-Colonialist Enterprise in Leslie Marmon Silko's *Almanac of the Dead*." *Modern Fiction Studies* 50, no. 2 (2004): 303–31.

Brookey, Robert Allen. *Reinventing the Male Homosexual: The Rhetoric and Power of the Gay Gene*. Bloomington: Indiana University Press, 2002.

Brotherston, Gordon. *Book of the Fourth World: Reading the Native Americas Through Their Literature*. New York: Cambridge University Press, 1992.

Brown, Phil. *Toxic Exposures: Contested Illnesses and the Environmental Health Movement*. New York: Columbia University Press, 2007.

Buell, Frederick. *From Apocalypse to Way of Life: Environmental Crisis in the American Century*. New York: Routledge, 2003.

Buell, Lawrence. "Ecocriticism: Some Emerging Trends." *Qui Parle* 19, no. 2 (2011): 87–115.

———. "Ecoglobalist Affects: The Emergence of U.S. Environmental Imagination on a Planetary Scale." In *Shades of the Planet: American Literature as World Literature*, ed. Wai Chee Dimock and Lawrence Buell, 227–48. Princeton: Princeton University Press, 2007.

———. *The Environmental Imagination: Thoreau, Nature Writing, and the Formation of American Culture*. Cambridge: Belknap Press of Harvard University Press, 1995.

———. *The Future of Environmental Criticism: Environmental Crisis and Literary Imagination*. Malden, MA: Blackwell, 2005.

———. *Writing for an Endangered World: Literature, Culture, and Environment in the U.S. and Beyond*. Cambridge: Belknap Press of Harvard University Press, 2001.

Burn, Stephen J. "'Webs of Nerves Pulsing and Firing': *Infinite Jest* and the Science of Mind." In *A Companion to David Foster Wallace Studies*, ed. Marshall Boswell and Stephen J. Burn, 59–85. New York: Palgrave Macmillan, 2013.

Bury, Liz. "Tugging at Heart Strings." *Bookseller*, 22 February 2007. Accessed 11 March 2012. http://www.thebookseller.com/feature/tugging-heart-strings.html.

Caldwell, Gail. "Awakenings." Review of *The Echo Maker*, by Richard Powers. *Boston Globe*, 8 October 2006.

Cameron, Sharon. *Writing Nature: Henry Thoreau's Journal*. New York: Oxford University Press, 1985.

Campbell, Sue. *Interpreting the Personal: Expression and the Formation of Feelings*. Ithaca, NY: Cornell University Press, 1997.

Carson, Rachel. *The Sense of Wonder*. New York: Harper and Row, 1965.

———. *Silent Spring*. Boston: Houghton Mifflin, 2002 [1962].

Carter, Erica, and Simon Watney, eds. *Taking Liberties: AIDS and Cultural Politics*. London: Serpent's Tale, 1989.

Chakrabarty, Dipesh. "The Climate of History: Four Theses." *Critical Inquiry* 35, no. 2 (2009): 197–222.

Chandler, David L. "An Earthbound Ark Sets Sail." *Boston Globe*, 20 August 1990.

Changeux, Jean-Pierre. *The Physiology of Truth: Neuroscience and Human Knowledge*. Trans. M. B. DeBevoise. Cambridge: Belknap Press of Harvard University Press, 2004.

Cheng, Anne Anlin. "Skins, Tattoos, and Susceptibility." *Representations* 108, no. 1 (2004): 98–119.

Clarke, Adele E., Laura Mamo, Jennifer Ruth Fosket, Jennifer R. Fishman, and Janet K. Shim, eds. *Biomedicalization: Technoscience, Health, and Illness in the U.S.* Durham, NC: Duke University Press, 2010.

Clough, Patricia T. "Introduction." In *The Affective Turn: Theorizing the Social*, ed. Patricia T. Clough and Jean Halley, 1–33. Durham, NC: Duke University Press, 2007.

Commoner, Barry. *The Closing Circle: Nature, Man, and Technology*. New York: Knopf, 1971.

Coole, Diana, and Samantha Frost, eds. *New Materialisms: Ontology, Agency, and Politics*. Durham, NC: Duke University Press, 2010.

Cooper, Marc. "Profits of Doom." *Village Voice*. 30 July 1991, 31–36.
Cooper, Melinda. *Life as Surplus: Biotechnology and Capitalism in the Neoliberal Era*. Seattle: University of Washington Press, 2008.
Corburn, Jason. *Street Science: Community Knowledge and Environmental Health Justice*. Cambridge: MIT Press, 2005.
Cousins, Mark. "The Ugly (Part 1)." *AA Files*. 28 (1994): 61–64.
Critical Art Ensemble. "Fear and Profit in the Fourth Domain." 2009. Accessed 25 March 2013. http://www.critical-art.net/genterra.html.
———. "GenTerra: Transgenic Solutions for a Greener World." 2003. Accessed 25 March 2013. http://critical-art.net/Original/genterra/genWeb.html.
Cronon, William. "Introduction: In Search of Nature." In *Uncommon Ground: Rethinking the Human Place in Nature*, ed. William Cronon, 23–56. New York: Norton, 1995.
———. "The Trouble with Wilderness: or, Getting Back to the Wrong Nature." In *Uncommon Ground: Rethinking the Human Place in Nature*, ed. William Cronon, 69–90. New York: Norton, 1995.
Cultural Cognition Project, The. "Home." Accessed 28 March 2013. http://www.culturalcognition.net/.
Cvetkovich, Ann. *An Archive of Feelings: Trauma, Sexuality, and Lesbian Public Cultures*. Durham, NC: Duke University Press, 2003.
———. *Depression: A Public Feeling*. Durham, NC: Duke University Press, 2012.
Damasio, Antonio R. *Descartes' Error: Emotion, Reason, and the Human Brain*. New York: Putnam, 1994.
———. *Looking for Spinoza: Joy, Sorrow, and the Feeling Brain*. New York: Harcourt, 2003.
Daston, Lorraine, and Katharine Park. *Wonders and the Order of Nature: 1150–1750*. New York: Zone, 1998.
Davila, Florangela. "Immigration Dispute Spawns Factions, Anger in Sierra Club." *Seattle Times*, 28 February 2004.
Dawkins, Richard. *Unweaving the Rainbow: Science, Delusion, and the Appetite for Wonder*. Boston: Houghton Mifflin, 1998.
de Certeau, Michel. *The Practice of Everyday Life*. 1980. Trans. Steven Rendall. Berkeley: University of California Press, 1984.
de Crèvecoeur, J. Hector St. John. *Letters from an American Farmer; and Sketches of Eighteenth-Century America*. New York: Penguin, 1981 [1782].
DeLillo, Don. *White Noise: Text and Criticism*. New York: Viking, 1998 [1985].
De Moor, Katrien. "The Doctor's Role of Witness and Companion: Medical and Literary Ethics of Care in AIDS Physicians' Memoirs." *Literature and Medicine* 22, no. 2 (2003): 208–29.

Deresiewicz, William. "Science Fiction." Review of *The Echo Maker*, by Richard Powers. *Nation*, 9 October 2006, 25–28.
Descartes, René. *The Passions of the Soul*. Trans. Stephen Voss. New York: Hackett, 1989 [1646].
Devall, Bill, and George Sessions. *Deep Ecology: Living as if Nature Mattered*. Salt Lake City: Smith, 1985.
Dewey, John. *Art as Experience*. New York: Penguin, 2005 [1934].
Dewey, Joseph. *In a Dark Time: The Apocalyptic Temper in the American Novel of the Nuclear Age*. West Lafayette, IN: Purdue University Press, 1990.
Di Chiro, Giovanna. "Local Actions, Global Visions: Remaking Environmental Expertise." *Frontiers* 18, no. 2 (1997): 203–31.
Diedrich, Lise. "AIDS and Its Treatments: Two Doctors' Narratives of Healing, Desire, and Belonging." *Journal of Medical Humanities* 26, no. 4 (2005): 237–57.
Dimock, Wai Chee, and Lawrence Buell, eds. *Shades of the Planet: American Literature as World Literature*. Princeton: Princeton University Press, 2007.
———, and Priscilla Wald. "Preface. Literature and Science: Cultural Forms, Conceptual Exchanges." *American Literature* 74, no. 4 (2002): 705–14.
Dorn, Michael, and Glenda Laws. "Social Theory, Body Politics, and Medical Geography: Extending Kearns's Invitation." *Professional Geography* 46, no. 1 (1994): 106–10.
Douglas, Mary. *Purity and Danger: An Analysis of Concepts of Pollution and Taboo*. London: Routledge, 2002 [1966].
Draaisma, Douwe. "Echos, Doubles, and Delusions: Capgras Syndrome in Science and Literature." *Style* 43, no. 3 (2009): 429–41.
Eagleton, Terry. *The Illusions of Postmodernism*. Malden, MA: Blackwell, 1996.
Easterlin, Nancy. "'Loving Ourselves Best of All': Ecocriticism and the Adapted Mind." *Mosaic* 37, no. 3 (September 2004): n.p.
Edelman, Gerald M. *A Universe of Consciousness: How Matter Becomes Imagination*. New York: Basic Books, 2000.
Ehrenreich, Barbara. *Bright-Sided: How Positive Thinking Is Undermining America*. New York: Picador, 2010.
Ehrlich, Paul R. *The Population Bomb*. Rivercity, MA: Rivercity, 1975 [1968].
Eiseley, Loren. *The Immense Journey: An Imaginative Naturalist Explores the Mysteries of Man and Nature*. New York: Vintage, 1973 [1957].
Empson, William. *Some Versions of Pastoral*. New York: New Directions, 1974.

Epstein, Julia. *Altered Conditions: Disease, Medicine, and Storytelling.* New York: Routledge, 1995.
Epstein, Steven. *Impure Science: AIDS, Activism, and the Politics of Knowledge.* Berkeley: University of California Press, 1996.
Eskridge, William N. *Dishonorable Passions: Sodomy Laws in America, 1861–2003.* New York: Penguin, 2008.
Estok, Simon C. "Reading Ecophobia: A Manifesto." *Ecozon@* 1, no. 1 (2010): 75–79.
———. "Theorizing in a Space of Ambivalent Openness: Ecocriticism and Ecophobia." *ISLE* 16, no. 2 (Spring 2009): 203–25.
Feagin, Susan L. "The Pleasures of Tragedy." *American Philosophical Quarterly* 20, no. 1 (1983): 95–104.
Feinberg, Todd E., and Julian Paul Keenan. "Where in the Brain Is the Self?" *Consciousness and Cognition* 14 (2005): 661–78.
Felski, Rita. "After Suspicion." *Profession* (2009): 28–35.
———. "Suspicious Minds." *Poetics Today* 32, no. 2 (2011): 215–34.
———. *Uses of Literature.* Malden, MA: Blackwell, 2008.
Fetterley, Judith, and Marjorie Pryse. *Writing out of Place: Regionalism, Women, and Literary Culture.* Urbana: University of Illinois Press, 2003.
Finkle, David. "Postcards from America." Review of *Close to the Knives: A Memoir of Disintegration*, by David Wojnarowicz. *Nation*, 26 August 1991, 238–39.
Fischer-Hornung, Dorothea. "Economies of Memory: Trafficking in Blood, Body Parts, and Crossblood Ancestors." *Amerikastudien/American Studies* 47, no. 2 (2002): 199–221.
Fisher, Philip. *Hard Facts: Setting and Form in the American Novel.* New York: Oxford University Press, 1985.
———. *The Vehement Passions.* Princeton: Princeton University Press, 2002.
———. *Wonder, the Rainbow, and the Aesthetics of Rare Experiences.* Cambridge: Harvard University Press, 1998.
Forster, E. M. *Howards End.* New York: Vintage, 1989 [1910].
Foucault, Michel. *The History of Sexuality:* vol. 1, *An Introduction.* Trans. Robert Hurley. New York: Vintage, 1990 [1976].
———. "The Masked Philosopher." In *Politics, Philosophy, Culture: Interviews and Other Writings, 1977–1984*, ed. Lawrence Kritzman, 323–30. New York: Routledge, 1990.
Freudenthal, Elizabeth. "Anti-Interiority: Compulsiveness, Objectification, and Identity in *Infinite Jest.*" *New Literary History* 41, no. 1 (Winter 2010): 191–211.

Fuller, Robert C. *Wonder: From Emotion to Spirituality*. Chapel Hill: University of North Carolina Press, 2006.

Furberg, Anne-Sofie, Torkjel Sandanger, Inger Thune, Ivan C. Burkow, and Eiliv Lund. "Fish Consumption and Plasma Levels of Organochlorines in a Female Population in Northern Norway." *Journal of Environmental Monitoring* 4, no. 1 (February 2002): 175–81.

Gallese, Vittorio. "'Being Like Me': Self-Other Identity, Mirror Neurons, and Empathy." In *Perspectives on Imitation: From Neuroscience to Social Science*, ed. Susan Hurley and Nick Chater, 1:101–18. Cambridge: MIT Press, 2005.

Garrard, Greg. "Ecocriticism." *Year's Work in Critical and Cultural Theory* 19 (2011): 46–82.

———. *Ecocriticism*. 2d ed. New York: Routledge, 2012.

———. "Nature Cures? or How to Police Analogies of Personal and Ecological Health." *ISLE* 19, no. 3 (Summer 2012): 494–514.

Gifford, Terry. *Pastoral*. New York: Routledge, 1999.

———. "Recent Critiques of Ecocriticism." *New Formations* 64 (2008): 15–24.

Glacken, Clarence J. *Traces on the Rhodian Shore: Nature and Culture in Western Thought from Ancient Times to the End of the Eighteenth Century*. Berkeley: University of California Press, 1967.

Glotfelty, Cheryll. "Literary Place Bashing, Test Site Nevada." In *Beyond Nature Writing: Expanding the Boundaries of Ecocriticism*, ed. Karla Armbruster and Kathleen R. Wallace, 233–47. Charlottesville: University of Virginia Press, 2001.

Gould, Stephen Jay. "The Terrifying Normalcy of AIDS." *New York Times Magazine*. 19 April 1987.

Gregg, Melissa, and Gregory J. Seigworth. "An Inventory of Shimmers." In *The Affect Theory Reader*, ed. Melissa Gregg and Gregory J. Seigworth, 1–28. Durham, NC: Duke University Press, 2010.

Grosz, Elizabeth. "Bodies-Cities." In *Places Through the Body*, ed. Heidi J. Nast and Steve Pile, 42–51. New York: Routledge, 1998.

Grover, Jan Zita. "Constitutional Symptoms." In *Taking Liberties: AIDS and Cultural Politics*, ed. Erica Carter and Simon Watney, 147–59. London: Serpent's Tale, 1989.

———. *North Enough: AIDS and Other Clear-Cuts*. Saint Paul, MN: Graywolf, 1997.

Guha, Ramachandra. *Environmentalism: A Global History*. New York: Longman, 2001.

———. "Radical American Environmentalism and Wilderness Preservation: A Third World Critique." *Environmental Ethics* 11 (Spring 1989): 71–83.

Hacking, Ian. "Our Neo-Cartesian Bodies in Parts." *Critical Inquiry* 34 (Autumn 2007): 78–105.

Halberstam, Judith. *In A Queer Time and Place: Transgender Bodies, Subcultural Lives.* New York: New York University Press, 2005.

Hale, Dorothy J. "Aesthetics and the New Ethics: Theorizing the Novel in the Twenty-First Century." *PMLA* 124, no. 3 (2009): 896–905.

Hamer, Dean. *The Science of Desire: The Search for the Gay Gene and the Biology of Behavior.* New York: Simon and Schuster, 1994.

Haraway, Donna. "The Promises of Monsters: A Regenerative Politics for Inappropriate/d Others." In *Cultural Studies*, ed. Lawrence Grossberg, Cary Nelson, and Paula A. Treichler, 295–337. New York: Routledge, 1992.

———. *Simians, Cyborgs, and Women: The Reinvention of Nature.* New York: Routledge, 1991.

Harré, Rom, Jens Brockmeier, and Peter Mülhaüser. *Greenspeak: A Study of Environmental Discourse.* Thousand Oaks, CA: Sage, 1999.

Harris, Charles B. "The Story of the Self: *The Echo Maker* and Neurological Realism." In *Intersections: Essays on Richard Powers*, ed. Stephen J. Burn and Peter Dempsey, 230–59. Champaign, IL: Dalkey Archive, 2008.

Harrison, Robert Pogue. *Gardens: An Essay on the Human Condition.* Chicago: University of Chicago Press, 2008.

Harvey, David. *A Brief History of Neoliberalism.* New York: Oxford University Press, 2005.

Hawken, Paul. *Blessed Unrest: How the Largest Social Movement in History Is Restoring Grace, Justice, and Beauty to the World.* New York: Penguin, 2008.

Hayles, N. Katherine. *How We Became Posthuman: Virtual Bodies in Cybernetics, Literature, and Informatics.* Chicago: University of Chicago Press, 1999.

———. "The Illusion of Autonomy and the Fact of Recursivity: Virtual Ecologies, Entertainment, and *Infinite Jest*." *New Literary History* 30, no. 3 (1999): 675–97.

Haynes, Todd, dir. *Safe.* 1995. Sony Pictures, 2001. DVD.

Hays, Samuel. *Beauty, Health, and Permanence: Environmental Politics in the United States, 1955–1985.* New York: Cambridge University Press, 1993 [1987].

Head, Dominic. "The (Im)possibility of Ecocriticism." In *Writing the Environment: Ecocriticism and Literature*, ed. Richard Kerridge and Neil Sammells, 27–39. London: Zed, 1998.

Heise, Ursula K. "The Hitchhiker's Guide to Ecocriticism." *PMLA* 121, no. 2 (2006): 503–16.

———. *Sense of Place and Sense of Planet: The Environmental Imagination of the Global.* New York: Oxford University Press, 2008.

Herman, Luc, and Bart Vervaeck. "Capturing Capgras: *The Echo Maker* by Richard Powers." *Style* 43, no. 3 (Fall 2009): 407–28.

Herndl, Carl G., and Stuart Cameron Brown, eds. *Green Culture: Environmental Rhetoric in Contemporary America.* Madison: University of Wisconsin Press, 1996.

Herring, Scott. *Another Country: Queer Anti-Urbanism.* New York: New York University Press, 2010.

Herzlich, Claudine. "Modern Medicine and the Quest for Meaning: Illness as a Social Signifier." In *The Meaning of Illness: Anthropology, History, and Sociology,* ed. Marc Augé and Claudine Herzlich, 151–73. Trans. Katherine J. Durnin, Caroline Lambein, Karen Leclercq-Jones, Barbara Puffer Garnier, and Ronald W. Williams. New York: Harwood Academic, 1995.

Hirstein, William, and V. S. Ramachandran. "Capgras Syndrome: A Novel Probe for Understanding the Neural Representation of the Identity and Familiarity of Persons." *Proceedings: Biological Sciences* 264, no. 1380 (1997): 437–44.

Hobbes, Thomas. *Leviathan.* 2d ed. Cambridge: Cambridge University Press, 1996 [1651].

Horvitz, Deborah. "Freud, Marx, and Chiapas in Leslie Marmon Silko's *Almanac of the Dead.*" *Studies in American Indian Literature* 10, no. 3 (1998): 47–64.

Howard, John. *Men Like That: A Southern Queer History.* Chicago: University of Chicago Press, 1999.

Huhndorf, Shari. *Mapping the Americas: The Transnational Politics of Contemporary Native Culture.* Ithaca, NY: Cornell University Press, 2009.

Hunt, Alex. "The Radical Geography of Silko's *Almanac of the Dead.*" *Western American Literature* 39, no. 3 (2004): 256–78.

Iacobini, Marco. *Mirroring People: The New Science of How We Connect with Others.* New York: Farrar, Straus and Giroux, 2008.

Illich, Ivan. *Medical Nemesis: The Expropriation of Health.* New York: Pantheon, 1976.

Indigenous Alliance of the Americas on 500 Years of Resistance. "Declaration of Quito." 1990. Accessed 20 May 2010. http://www.nativeweb.org/papers/statements/quincentennial/quito.php#declaration.

International Covenant on Economic, Cultural, and Social Rights. United Nations. 19 December 1966, 933 *U.N.T.S.* 3.

International Union for the Conservation of Nature. "Summary Statistics for Globally Threatened Species. Table 1: Numbers of Threatened Species by Major Groups of Organisms (1996–2012)." *IUCN.org*. 2012. Accessed 31 August 2013.

Irr, Caren. "The Timeliness of *Almanac of the Dead*, or a Postmodern Rewriting of Radical Fiction." In *Leslie Marmon Silko: A Collection of Critical Essays*, ed. Louise K. Barnett and James L. Thorson, 223–44. Albuquerque: University of New Mexico Press, 1999.

Izard, Carroll, and Brian Ackerman. "Motivational, Organizational, and Regulatory Functions of Discrete Emotions." *The Handbook of Emotions*, ed. Michael Lewis and Jeannette Haviland-Jones, 253–64. 2d ed. New York: Guilford, 2004.

Jameson, Fredric. *Archaeologies of the Future: The Desire Called Utopia and Other Science Fictions*. New York: Verso, 2005.

———. *Postmodernism, or, the Cultural Logic of Late Capitalism*. Durham, NC: Duke University Press, 1991.

———. "Reading *Capital*." *Mediations* 25, no. 1 (Fall 2010): 5–14.

Jamieson, Dale. *Ethics and the Environment: An Introduction*. Cambridge: Cambridge University Press, 2008.

Jarman, Michelle. "Exploring the World of the Different in Leslie Marmon Silko's *Almanac of the Dead*." *MELUS* 31, no. 3 (2006): 147–68.

Jasanoff, Sheila. "Heaven and Earth: The Politics of Environmental Images." In *Earthly Politics: Local and Global in Environmental Governance*, ed. Sheila Jasanoff and Marybeth Long Martello, 31–54. Cambridge: MIT Press, 2004.

Jenkins, McKay. *What's Gotten Into Us? Staying Healthy in a Toxic World*. New York: Random House, 2011.

Johnson, E. Patrick. *Sweet Tea: Black Gay Men of the South*. Chapel Hill: University of North Carolina Press, 2008.

Jones, Malcolm. "Reports from the Heartland." Review of *Almanac of the Dead*, by Leslie Marmon Silko. *Newsweek*, 18 November 1991, 81.

Kahan, Dan M., Donald Braman, Paul Slovic, John Gastil, and Geoffrey L. Cohen. "The Second National Risk and Culture Study: Making Sense of—and Making Progress in—the American Culture War of Fact." *GWU Legal Studies Research Paper* no. 370 (2007). 26 September 2007. http://ssrn.com/abstract=1017189.

Kakutani, Michiko. "A Country Dying of Laughter: In 1,079 Pages." Review of *Infinite Jest*, by David Foster Wallace. *New York Times*, 13 February 1996.

———. "Imaginary Lives, Built in Empty Rooms." Review of *Plowing the Dark*, by Richard Powers. *New York Times*, 20 June 2000.
Karem, Jeff. *The Romance of Authenticity: The Cultural Politics of Regional and Ethnic Literatures*. Charlottesville: University of Virginia Press, 2004.
Keen, Suzanne. *Empathy and the Novel*. New York: Oxford University Press, 2007.
Kelleher, Kathleen. "Predicting a Revolt to Reclaim the Americas." *Los Angeles Times*, 13 January 1992.
Keller, Evelyn Fox. "Nature, Nurture, and the Human Genome Project." In *The Code of Codes: Scientific and Social Issues in the Human Genome Project*, ed. Daniel J. Kevles and Leroy Hood, 281–99. Cambridge: Harvard University Press, 1992.
———. *Refiguring Life: Metaphors of Twentieth-Century Biology*. New York: Columbia University Press, 1995.
Kevles, Daniel J. "Out of Genetics: The Historical Politics of the Human Genome." In *The Code of Codes: Scientific and Social Issues in the Human Genome Project*, ed. Daniel J. Kevles and Leroy Hood, 3–36. Cambridge: Harvard University Press, 1992.
Kierkegaard, Søren. *The Concept of Anxiety: A Simple Psychologically Orienting Deliberation on the Dogmatic Issue of Hereditary Sin*. Ed. and trans. Reidar Thomte. Princeton: Princeton University Press, 1980 [1844].
Killingsworth, M. Jimmie, and Jacqueline S. Palmer. "Millennial Ecology: The Apocalyptic Narrative from *Silent Spring* to Global Warming." In *Green Culture: Environmental Rhetoric in Contemporary America*, ed. Carl G. Herndl and Stuart Cameron Brown, 20–45. Madison: University of Wisconsin Press, 1996.
Kleinman, Arthur. *The Illness Narratives: Suffering, Healing, and the Human Condition*. New York: Basic Books, 1988.
Kolodny, Annette. *The Lay of the Land: Metaphor as Experience and History in American Life and Letters*. Chapel Hill: University of North Carolina Press, 1975.
Knickerbocker, Brad. "Dark Beauty, Bright Terror." Review of *Almanac of the Dead*, by Leslie Marmon Silko. *Christian Science Monitor*, 3 February 1992.
———. "A 'Hostile' Takeover Bid at the Sierra Club." *Christian Science Monitor*, 20 February 2004.
Kristeva, Julia. *Powers of Horror: An Essay on Abjection*. Trans. Leon S. Roudiez. New York: Columbia University Press, 1982.

Kroll-Smith, Steven, and H. Hugh Floyd. *Bodies in Protest: Environmental Illness and the Struggle Over Medical Knowledge*. New York: New York University Press, 1997.

Krupat, Arnold. *The Turn to the Native: Studies in Criticism and Culture*. Lincoln: University of Nebraska Press, 1996.

Kuzyk, Yuri. "An Excellent Book...," Customer review of *Almanac of the Dead*, by Leslie Marmon Silko. Amazon.com. 4 September 2000. Accessed 2 April 2012. http://www.amazon.com/review/R2EL7T291SA5WS/ref=cm_cr_pr_perm?ie=UTF8&ASIN=0140173196&linkCode=&nodeID=&tag=.

Ladino, Jennifer K. *Reclaiming Nostalgia: Longing for Nature in American Literature*. Charlottesville: University of Virginia Press, 2012.

Langston, Nancy. *Toxic Bodies: Hormone Disruptors and the Legacy of DES*. New Haven: Yale University Press, 2010.

Latour, Bruno. "'It's the Development, Stupid!' Or How to Modernize Modernization." Review of *Break Through: From the Death of Environmentalism to the Politics of Possibility*, by Ted Nordhaus and Michael Shellenberger. *Espacestemps.net*, 29 May 2008. http://www.espacestemps.net/en/articles/itrsquos-development-stupid-or-how-to-modernize-modernization-en/.

———. "Why Has Critique Run out of Steam? From Matters of Fact to Matters of Concern." *Critical Inquiry* 30, no. 2 (Winter 2004): 225–48.

Lawrence v. Texas. 539 U.S. 538 (2003).

LeClair, Tom. "The Prodigious Fiction of Richard Powers, William Vollmann, and David Foster Wallace." *Critique* 38, no. 1 (1996): 12–37.

LeDoux, Joseph E. *Synaptic Self: How Our Brains Become Who We Are*. New York: Penguin, 2002.

Lefebvre, Henri. *The Production of Space*. Trans. Donald Nicholson-Smith. Cambridge: Blackwell, 1991 [1974].

Leopold, Aldo. *Round River: From the Journals of Aldo Leopold*, ed. Luna B. Leopold. New York: Oxford University Press, 1993 [1953].

———. *A Sand County Almanac, with Essays on Conservation from Round River*. New York: Ballantine, 1966 [1949].

Le Sueur, Meridel. "Eroded Woman." 1948. In *Harvest and Song for My Time: Stories*, 83–89. Minneapolis: West End Press and MEP, 1982.

LeVay, Simon. "A Difference in Hypothalamic Structure Between Heterosexual and Homosexual Men." *Science* 253, no. 5023 (30 August 1991): 1034–37.

———. *The Sexual Brain*. Cambridge: MIT Press, 1993.

Lew, Julie. "Peering Into a World Under Glass." *New York Times*, 22 December 1991.
Leys, Ruth. "The Turn to Affect: A Critique." *Critical Inquiry* 37, no. 3 (2011): 434–72.
Limerick, Patricia Nelson. *Something in the Soil: Legacies and Reckonings in the New West*. New York: Norton, 2000.
Lipsky, David. *Although Of Course You End Up Becoming Yourself: A Road Trip with David Foster Wallace*. New York: Broadway, 2010.
Love, Heather. "Close but Not Deep: Literary Ethics and the Descriptive Turn." *New Literary History* 41, no. 2 (2010): 371–91.
Ludden, Jennifer. "Ads Warn That All Immigration Must Be Reduced." *NPR.org. Morning Edition*. 12 September 2008. Accessed 1 April 2012. http://www.npr.org/templates/story/story.php?storyId=94545604.
Lynch, Tom. *Xerophilia: Ecocritical Explorations in Southwestern Literature*. Lubbock: Texas Tech University Press, 2008.
Manovich, Lev. "Database as Symbolic Form." In *Database Aesthetics: Art in the Age of Information Overflow*, ed. Victoria Vesna, 39–60. Minneapolis: University of Minnesota Press, 2007.
Marsh, George Perkins. *Man and Nature, or, Physical Geography as Modified by Human Action*, ed. David Lowenthal. Seattle: University of Washington Press, 2003 [1864].
Marx, Leo. *The Machine in the Garden: Technology and the Pastoral Ideal in America*. London: Oxford University Press, 1967.
Massumi, Brian. *Parables for the Virtual: Movement, Affect, Sensation*. Durham, NC: Duke University Press, 2002.
———. "Preface." In *The Politics of Everyday Fear*, vii–x. Minneapolis: University of Minnesota Press, 1993.
Max, D. T. *Every Love Story Is a Ghost Story: A Life of David Foster Wallace*. New York: Viking, 2012.
McKibben, Bill. *The End of Nature*. New York: Random House, 2006 [1989].
McNeill, J. R. *Something New Under the Sun: An Environmental History of the Twentieth-Century World*. New York: Norton, 2000.
Merchant, Carolyn. *American Environmental History: An Introduction*. New York: Columbia University Press, 2007.
Metzl, Jonathan M., and Anna Kirkland, eds. *Against Health: How Health Became the New Morality*. New York: New York University Press, 2010.
Michaels, Walter Benn. *The Shape of the Signifier: 1967 to the End of History*. Princeton: Princeton University Press, 2004.

Miller, D. A. *Jane Austen, or The Secret of Style*. Princeton: Princeton University Press, 2003.

Miller, William Ian. *The Anatomy of Disgust*. Cambridge: Harvard University Press, 1997.

Mims, Christopher. "80 Percent of Humans Are Delusionally Optimistic, Says Science." Grist.org. 10 April 2012. http://grist.org/list/80-percent-of-humans-are-delusionally-optimistic-says-science/.

Mitman, Gregg. *Breathing Space: How Allergies Shape Our Lives and Landscapes*. New Haven: Yale University Press, 2007.

Mittal, Anuradha, and Mayumi Kawaai. "Freedom to Trade? Trading Away American Family Farms." *Food First Backgrounder* 7, no. 4 (2001). Accessed 31 August 2013. http://www.foodfirst.org/en/node/56.

Miyoshi, Masao. "Turn to the Planet: Literature, Diversity, and Totality." *Comparative Literature* 53, no. 4 (Autumn 2001): 283–97.

Mogen, David. "Native American Visions of Apocalypse: Prophecy and Protest in the Fiction of Leslie Marmon Silko and Gerald Vizenor." In *American Mythologies: Essays on Contemporary Literature*, ed. William Blazek and Michael K. Glenday, 157–67. Liverpool: Liverpool University Press, 2005.

Montrie, Chad. *A People's History of Environmentalism in the United States*. New York: Continuum, 2011.

Moore, David L. "Silko's Blood Sacrifice: The Circulating Witness in *Almanac of the Dead*." In *Leslie Marmon Silko: A Collection of Critical Essays*, ed. Louise K. Barnett and James L. Thorson, 149–83. Albuquerque: University of New Mexico Press, 1999.

Moretti, Franco. *Graphs, Maps, Trees: Abstract Models for a Literary History*. New York: Verso, 2005.

Mortimer-Sandilands, Catriona. "'I Still Need the Revolution': Cultivating Ecofeminist Readers." In *Teaching North American Environmental Literature*, ed. Laird Christensen, Mark C. Long, and Fred Waage, 58–71. New York: Modern Language Association of America, 2008.

———, and Bruce Erickson, eds. *Queer Ecologies: Sex, Nature, Politics, Desire*. Bloomington: Indiana University Press, 2010.

Morton, Timothy. *The Ecological Thought*. Cambridge: Harvard University Press, 2010.

———. *Ecology Without Nature: Rethinking Environmental Aesthetics*. Cambridge: Harvard University Press, 2007.

Muir, John. *My First Summer in the Sierra*. Boston: Houghton Mifflin, 1911.

Mumford, Lewis. *The City in History: Its Origins, Its Transformations, and Its Prospects*. Boston: Houghton Mifflin, 1961.

Muñoz, José Esteban. *Cruising Utopia: The Then and There of Queer Futurity*. New York: New York University Press, 2009.

Nash, Linda Lorraine. *Inescapable Ecologies: A History of Environment, Disease, and Knowledge*. Berkeley: University of California Press, 2006.

Ngai, Sianne. *Ugly Feelings*. Cambridge: Harvard University Press, 2005.

Nichols, Catherine. "Dialogizing Postmodern Carnival: David Foster Wallace's *Infinite Jest*." *Critique* 43, no. 1 (2001): 3–16.

Nietzsche, Friedrich. *The Gay Science*. Trans. Walter Kaufmann. New York: Vintage, 1974 [1887].

Nixon, Rob. *Slow Violence and the Environmentalism of the Poor*. Cambridge: Harvard University Press, 2011.

"No Compromise in Defense of Mother Earth." *EarthFirstJournal.org*. Earth First Journal! Accessed 1 April 2012. http://earthfirstjournal.org/section.php?id=1.

Norden, Christopher. "Ecological Restoration as Post-Colonial Ritual of Community in Three Native American Novels." *Studies in American Indian Literature* 6, no. 4 (1994): 94–106.

Oatley, Keith. *Best Laid Schemes: The Psychology of Emotions*. New York: Oxford University Press, 1992.

Olmsted, Jane. "The Uses of Blood in Leslie Marmon Silko's *Almanac of the Dead*." *Contemporary Literature* 40, no. 3 (1999): 464–90.

O'Meara, Bridget. "The Ecological Politics of Leslie Silko's *Almanac of the Dead*." *Wicazo Sa Review* 15, no. 2 (2000): 63–73.

O'Neill, Robert V. "Is It Time to Bury the Ecosystem Concept (with Full Military Honors, of Course)?" *Ecology* 82, no. 12 (2001): 3275–84.

Oppermann, Serpil. "Ecocriticism's Theoretical Discontents." *Mosaic* 44, no. 2 (2011): 153–69.

Packard, Vance. *The Waste Makers*. New York: David McKay, 1960.

Palumbo-Liu, David. *The Deliverance of Others: Reading Literature in a Global Age*. Durham, NC: Duke University Press, 2012.

Patton, Cindy. *Inventing AIDS*. New York: Routledge, 1990.

Pellegrini, Ann, and Jasbir Puar. "Affect." *Social Text* 27, no. 3 (2009): 35–38.

Perrow, Charles. *Normal Accidents: Living with High-Risk Technologies*. New York: Basic Books, 1984.

Phillips, Dana. *The Truth of Ecology: Nature, Culture, and Literature in America*. New York: Oxford University Press, 2003.

Phillips, Richard, Diane Watt, and David Shuttleton. *De-Centering Sexualities: Politics and Representations Beyond the Metropolis*. New York: Routledge, 1999.

Piercy, Marge. "The City as Battleground: The Novelist as Combatant." In *Literature and the Urban Experience: Essays on the City and Literature*, ed. Michael C. Jaye and Ann Chalmers Watts, 209–17. New Brunswick, NJ: Rutgers University Press, 1981.

———. *Woman on the Edge of Time*. New York: Ballantine, 1983 [1976].

"Plane Stupid." *PlaneStupid.org*. Plane Stupid. Accessed 2 April 2012. http://www.planestupid.com/polarbears.

Pole, David. "Disgust and Other Forms of Aversion." In *Aesthetics, Form, and Emotion*, ed. George Roberts, 219–31. New York: St. Martin's, 1983.

Powers, Janet M. "Mapping the Prophetic Landscape in *Almanac of the Dead*." In *Leslie Marmon Silko: A Collection of Critical Essays*, ed. Louise K. Barnett and James L. Thorson, 261–72. Albuquerque: University of New Mexico Press, 1999.

Powers, Richard. "The Art of Fiction." Interview by Kevin Berger. *Paris Review* 164, no. 3 (2002): 106–38.

———. *The Echo Maker*. New York: Picador, 2006.

———. "Environmental Writing in Four Dimensions: Fiction." Lecture given at Environmental Humanities Project, Stanford University, Stanford, CA, 5 March 2010.

———. *Gain*. New York: Picador, 1998.

———. *Galatea 2.2: A Novel*. New York: HarperPerennial, 1995.

———. *The Gold Bug Variations*. New York: HarperPerennial, 1991.

———. "An Interview with Richard Powers." Interview by Randall Fuller. *Missouri Review* 26, no. 1 (2003): 99–112.

———. "An Interview with Richard Powers." Interview by Stephen J. Burn. *Contemporary Literature* 49, no. 2 (2008): 163–79.

———. "The Last Generalist." Interview by Jeffrey Williams. *Clogic.eserver.org*. *Cultural Logic* 2, no. 2 (1999).

———. "Making the Rounds." In *Intersections: Essays on Richard Powers*, ed. Stephen J. Burn and Peter Dempsey, 305–10. Champaign, IL: Dalkey Archive, 2008.

———. "What Does Fiction Know?" *Places*. 2 August 2011. http://places.designobserver.com/feature/what-does-fiction-know-richard-powers/28838/.

Probyn, Elspeth. "Writing Shame." In *The Affect Theory Reader*, ed. Melissa Gregg and Gregory J. Seigworth, 71–90. Durham, NC: Duke University Press, 2010.

Prüss-Üstün, Annette, and Carlos Corvalán. "Preventing Disease Through Healthy Environments: Towards an Estimate of the Environmental Burden of Disease." Geneva: World Health Organization Press, 2006. Accessed 12 March 2012. http://www.who.int/quantifying_ehimpacts/publications/preventingdisease.pdf

Rabinow, Paul. "Artificiality and Enlightenment: From Sociobiology to Biosociality." In *Incorporations*, ed. Jonathan Crary and Sanford Kwinter, 234–52. New York: Zone, 1992.

———, and Nikolas Rose. "Biopower Today." *Biosocieties* 1 (2006): 195–217.

Rajan, Kaushik Sunder. *Biocapital: The Constitution of Postgenomic Life*. Durham, NC: Duke University Press, 2006.

Ramachandran, V. S., and Sandra Blakeslee. *Phantoms in the Brain: Probing the Mysteries of the Human Mind*. New York: William Morrow, 1998.

Ramlow, Todd R. "Bodies in the Borderlands: Gloria Anzaldúa's and David Wojnarowicz's Mobility Machines." *MELUS* 31, no. 3 (2006): 169–87.

Reed, T. V. "Toxic Colonialism, Environmental Justice, and Native Resistance in Silko's *Almanac of the Dead*." *MELUS* 34, no. 2 (2009): 25–42.

Reineke, Yvonne. "Overturning the (New World) Order: Of Space, Time, Writing, and Prophecy in Leslie Marmon Silko's *Almanac of the Dead*." *Studies in American Indian Literature* 10, no. 3 (1998): 65–83.

Ricoeur, Paul. *Freud and Philosophy: An Essay on Interpretation*. Trans. Denis Savage. New Haven: Yale University Press, 1970.

———. *Time and Narrative*, vol. 1. Trans. Kathleen McLaughlin and David Pellauer. Chicago: University of Chicago Press, 1984.

Rizzolatti, Giacomo, and Corrado Sinigaglia. *Mirrors in the Brain: How Our Minds Share Actions and Emotions*. Trans. Frances Anderson. Oxford: Oxford University Press, 2008.

Robinson, Douglas. *American Apocalypses: The Image of the End of the World in American Literature*. Baltimore: Johns Hopkins University Press, 1985.

Robisch, S. K. "The Woodshed: A Response to 'Ecocriticism and Ecophobia.'" *ISLE* 16, no. 4 (Autumn 2009): 697–708.

Romero, Channette. "Envisioning a 'Network of Tribal Coalitions': Leslie Marmon Silko's *Almanac of the Dead*." *American Indian Quarterly* 26, no. 4 (2002): 623–40.

Rorty, Amélie. "From Passions to Emotions and Sentiments." *Philosophy* 57, no. 220 (1982): 159–72.

Rose, Hilary, and Steven Rose. *Genes, Cells, and Brains: The Promethean Promises of the New Biology*. London: Verso, 2012.

Rose, Nikolas. *The Politics of Life Itself: Biomedicine, Power, and Subjectivity in the Twenty-First Century*. Princeton: Princeton University Press, 2007.

Ross, Andrew. *Bird on Fire: Lessons from the World's Least Sustainable City*. New York: Oxford University Press, 2011.

———. *The Chicago Gangster Theory of Life: Nature's Debt to Society*. New York: Verso, 1994.

Rotello, Gabriel. *Sexual Ecology: AIDS and the Destiny of Gay Men*. New York: Plume, 1997.

Rozin, Paul, Jonathan Haidt, and Clark R. McCauley. "Disgust." In *Handbook of Emotions*, ed. Michael Lewis, Jeannette Haviland-Jones, and Lisa Feldman Barrett, 757–76. 3rd ed. New York: Guilford, 2008.

Russell, Emily. "Some Assembly Required: The Embodied Politics of *Infinite Jest*." *Arizona Quarterly* 66, no. 2 (2010): 147–69.

Sadowski-Smith, Claudia. *Border Fictions: Globalization, Empire, and Writing at the Boundaries of the United States*. Charlottesville: University of Virginia Press, 2008.

Schipper, Jan, Janice S. Chanson, Federica Chiozza, Neil A. Cox et al. "The Status of the World's Land and Marine Mammals: Diversity, Threat, and Knowledge." *Science* 322, no. 5899 (2008): 225–30.

Schlosser, Eric. *Fast Food Nation: The Dark Side of the All-American Meal*. New York: Houghton Mifflin, 2001.

Sedgwick, Eve Kosofsky. *Touching Feeling: Affect, Pedagogy, Performativity*. Durham, NC: Duke University Press, 2003.

Sennett, Richard. *Flesh and Stone: The Body and the City in Western Civilization*. New York: Norton, 1994.

Serres, Michel. *The Natural Contract*. Trans. Elizabeth MacArthur and William Paulson. Ann Arbor: University of Michigan Press, 1995.

Sharot, Tali, Christoph W. Korn, and Raymond J. Dolan. "How Unrealistic Optimism Is Maintained in the Face of Reality." *Nature Neuroscience* 14, no. 11 (2011): 1475–79.

Shilts, Randy. *And the Band Played On: Politics, People, and the AIDS Epidemic*. New York: St. Martin's Griffin, 2007 [1987].

Shklovsky, Viktor. "Art as Technique." *Russian Formalist Criticism: Four Essays*. Trans. Lee T. Lemon and Marion J. Reis, 3–24. Lincoln: University of Nebraska Press, 1965 [1917].

Silko, Leslie Marmon. *Almanac of the Dead: A Novel*. New York: Penguin, 1991.

———. *Ceremony*. New York: Penguin, 2006 [1977].

———. "The Fourth World." *Artforum* 27, no. 10 (1989): 124–27.
———. "An Interview with Leslie Marmon Silko." Interview by Florence Boos. In *Conversations with Leslie Marmon Silko*, ed. Laura Coltelli, 135–45. Jackson: University of Mississippi Press, 2000 [1994].
———. *Yellow Woman and a Beauty of the Spirit: Essays on Native American Life Today*. New York: Simon and Schuster, 1996.
Simmel, Georg. "The Metropolis and Mental Life." 1903. In *The Blackwell City Reader*, ed. Gary Bridge and Sophie Watson, 11–19. Malden, MA: Blackwell, 2002.
Sinclair, Upton. *The Jungle*. New York: Penguin, 2006 [1906].
Slovic, Paul, and Scott Slovic. "Numbers and Nerves: Toward an Effective Apprehension of Environmental Risk." *Whole Terrain* 13 (2004/2005): 14–18.
Slovic, Scott. "Nature Writing." In *Encyclopedia of World Environmental History*, ed. Shepard Krech III, J. R. McNeill, and Carolyn Merchant, 2:886–91. New York: Routledge, 2004.
———. *Seeking Awareness in American Nature Writing: Henry Thoreau, Annie Dillard, Edward Abbey, Wendell Berry, Barry Lopez*. Salt Lake City: University of Utah Press, 1992.
Smalley, E. V. "The Isolation of Life on Prairie Farms." *Atlantic Monthly* 72, no. 431 (1893): 378–82.
Smith, Marquard, and Joanne Morra, eds. *The Prosthetic Impulse: From a Posthuman Present to a Biocultural Future*. Cambridge: MIT Press, 2006.
Smith, Neil. *Uneven Development: Nature, Capital, and the Production of Space*. Athens: University of Georgia Press, 2008 [1984].
Smith, Rachel Greenwald. "Postmodernism and the Affective Turn." *Twentieth-Century Literature* 57, no. 3–4 (Fall/Winter 2011): 423–46.
Soja, Edward W. *Postmodern Geographies: The Reassertion of Space in Critical Social Theory*. New York: Verso, 1989.
Solnit, Rebecca. *Hope in the Dark: Untold Histories, Wild Possibilities*. New York: Nation, 2004.
Sontag, Susan. *Illness as Metaphor and AIDS and Its Metaphors*. New York: Anchor, 1990.
Soper, Kate. *What Is Nature? Culture, Politics and the Non-Human*. Cambridge: Blackwell, 1995.
Speth, James Gustave. *America the Possible: Manifesto for a New Economy*. New Haven: Yale University Press, 2012.
Spivak, Gayatri Chakravorty. *Death of a Discipline*. New York: Columbia University Press, 2003.

Squier, Susan Merrill. *Liminal Lives: Imagining the Human at the Frontiers of Biomedicine.* Durham, NC: Duke University Press, 2004.

Stanford, Ann Folwell. "'Human Debris': Border Politics, Body Parts, and the Reclamation of the Americas in Leslie Marmon Silko's *Almanac of the Dead*." *Literature and Medicine* 16, no. 1 (1997): 23–42.

Starr, Kevin. *California: A History.* New York: Random House, 2005.

Stein, Rachel. *Shifting the Ground: American Women Writers' Revisions of Nature, Gender, and Race.* Charlottesville: University of Virginia Press, 1997.

Steingraber, Sandra. *Living Downstream: A Scientist's Personal Investigation of Cancer and the Environment.* New York: Vintage, 1998.

St. John, Edward B. "Almanac of the Dead." Review of *Almanac of the Dead*, by Leslie Marmon Silko. *Library Journal*, 15 October 1991.

Suplee, Curt. "Brave Small World." *Washington Post*, 21 January 1990.

Sutherland, John. "Paper or Plastic?" Review of *Gain* and *Plowing the Dark*, by Richard Powers. *London Review of Books*, 10 August 2000, 20–21.

Tabbi, Joseph. *Cognitive Fictions.* Minneapolis: University of Minnesota Press, 2002.

Tallent, Elizabeth. "Storytelling with a Vengeance." Review of *Almanac of the Dead*, by Leslie Marmon Silko. *New York Times*, 22 December 1991.

Taylor, Matthew A. "The Nature of Fear: Edgar Allan Poe and Posthuman Ecology." *American Literature* 84, no. 2 (June 2012): 353–79.

Teale, Tamara M. "The Silko Road from Chiapas or Why Native Americans Cannot Be Marxists." *MELUS* 23, no. 4 (1998): 157–66.

Thacker, Eugene. *Global Genome: Biotechnology, Politics, and Culture.* Cambridge: MIT Press, 2005.

Thomashow, Mitchell. *Bringing the Biosphere Home: Learning to Perceive Global Environmental Change.* Cambridge: MIT Press, 2002.

Thoreau, Henry David. *Journal*, vol. 4, ed. John C. Broderick and Robert Sattelmeyer. 1851–52. Princeton: Princeton University Press, 1981.

Thornber, Karen Laura. *Ecoambiguity: Environmental Crises and East Asian Literatures.* Ann Arbor: University of Michigan Press, 2012.

Thrailkill, Jane F. *Affecting Fictions: Mind, Body, and Emotion in American Literary Realism.* Cambridge: Harvard University Press, 2007.

Tillett, Rebecca. "Reality Consumed by Realty: The Ecological Costs of 'Development' in Leslie Marmon Silko's *Almanac of the Dead*." *European Journal of American Culture* 24, no. 2 (2005): 153–69.

Toufexis, Anastasia. "The Wizards of Hokum." *Time* 138, no. 13 (1991): 66.

Treichler, Paula A. "AIDS, Homophobia, and Biomedical Discourse: An Epidemic of Signification." *Cultural Studies* 1, no. 3 (1987): 263–305.

Tuan, Yi-fu. *Topophilia: A Study of Environmental Perception, Attitudes, and Values*. Englewood Cliffs, NJ: Prentice-Hall, 1974.

Turney, Jon. *Frankenstein's Footsteps: Science, Genetics and Popular Culture*. New Haven: Yale University Press, 1998.

Van Dyke, Annette. "From Big Green Fly to the Stone Serpent: Following the Dark Vision in Silko's *Almanac of the Dead*." *Studies in American Indian Literature* 10, no. 3 (1998): 34–46.

Verghese, Abraham. *My Own Country: A Doctor's Story*. New York: Vintage, 1994.

Verghese, Abraham, Steven Berk, and Felix Sarubbi. "*Urbs in Rure*: Human Immunodeficiency Virus Infection in Rural Tennessee." *Journal of Infectious Diseases* 160, no. 6 (1989): 1051–55.

Wald, Priscilla. "American Studies and the Politics of Life." *American Quarterly* 64, no. 2 (June 2012): 185–204.

———. *Contagious: Cultures, Carriers, and the Outbreak Narrative*. Durham, NC: Duke University Press, 2008.

Waldby, Catherine. *AIDS and the Body Politic: Biomedicine and Sexual Difference*. New York: Routledge, 1996.

———. *The Visible Human Project: Informatic Bodies and Posthuman Medicine*. New York: Routledge, 2000.

———, and Robert Mitchell. *Tissue Economies: Blood, Organs, and Cell Lines in Late Capitalism*. Durham, NC: Duke University Press, 2006.

Walker, Martin. "Home, Sweet Biome." *Guardian (London)*, 25 May 1990.

Wallace, David Foster. *Brief Interviews with Hideous Men*. New York: Back Bay, 2007 [1999].

———. *The Broom of the System*. New York: Penguin, 1987.

———. *Consider the Lobster and Other Essays*. New York: Little, Brown, 2006.

———. *Girl with Curious Hair*. New York: Norton, 1989.

———. *Infinite Jest*. New York: Back Bay, 1996.

———. "Interview with David Foster Wallace." Interview by Larry McCaffery. *Review of Contemporary Fiction* 13, no. 2 (1993): 127–50.

———. *A Supposedly Fun Thing I'll Never Do Again: Essays and Arguments*. New York: Back Bay, 1997.

———. *This Is Water: Some Thoughts, Delivered on a Significant Occasion, About Living a Compassionate Life*. New York: Little, Brown, 2009.

Wallerstein, Immanuel. *World-Systems Analysis: An Introduction*. Durham, NC: Duke University Press, 2004.

Ward, Bud. "About Us." *The Yale Forum on Climate Change & the Media.* Yale University. 2012. Accessed 11 March 2012. http://www.yaleclimatemediaforum.org/aboutus/.

———. "NOAA's 2012 Arctic 'Report Card'... Danger Signs Ahead." *Yale Forum on Climate Change & the Media.* Yale University. 6 December 2012. http://www.yaleclimatemediaforum.org/2012/12/noaas-2012-arctic-report-card-danger-signs-ahead/.

Westling, Louise Hutchings. *The Green Breast of the New World: Landscape, Gender, and American Fiction.* Athens: University of Georgia Press, 1996.

Weston, Kath. "Get Thee to a Big City: Sexual Imaginary and the Great Gay Migration." *GLQ* 2, no. 3 (1995): 253–77.

Wilde, Oscar. "The Decay of Lying: An Observation." In *The Decay of Lying and Other Essays.* London: Penguin, 2010 [1889].

Williams, Raymond. *The Country and the City.* New York: Oxford University Press, 1973.

———. "Ideas of Nature." In *Culture and Materialism: Selected Essays*, 67–85. London: Verso, 2005 [1972].

———. *Marxism and Literature.* New York: Oxford University Press, 1977.

Williams, Terry Tempest. *Refuge: An Unnatural History of Family and Place.* New York: Vintage, 1992.

Wohlforth, Charles. *The Fate of Nature: Rediscovering Our Ability to Rescue the Earth.* New York: Thomas Dunne, 2010.

Wojnarowicz, David. *Close to the Knives: A Memoir of Disintegration.* New York: Vintage, 1991.

———, and James Romberger. *Seven Miles a Second.* New York: DC Comics, 1996.

Wojnarowicz v. American Family Association. 745 F. Supp. 130 (S.D. NY 1990).

Wolfe, Cary. *What Is Posthumanism?* Minneapolis: University of Minnesota Press, 2010.

Womack, Craig. "Suspicioning: Imagining a Debate Between Those Who Get Confused, and Those Who Don't, When They Read Critical Responses to the Poems of Joy Harjo, or What's an Old-Timey Gay Boy Like Me to Do?" *GLQ* 16, no. 1–2 (2010): 133–55.

Worster, Donald. *Nature's Economy: A History of Ecological Ideas.* 2d ed. New York: Cambridge University Press, 1994 [1977].

Yoon, Carol Kaesuk. "Luminous 3-D Jungle Is a Biologist's Dream." *New York Times*, 19 January 2010.

Zamora, Lois Parkinson. *Writing the Apocalypse: Historical Vision in Contemporary U.S. and Latin American Fiction.* New York: Cambridge University Press, 1989.

Zuniga, José M., Alan Whiteside, Amin Ghaziani, and John G. Bartlett, eds. *A Decade of HAART: The Development and Global Impact of Highly Active Antiretroviral Therapy.* New York: Oxford University Press, 2008.

Zunshine, Lisa. *Strange Concepts and the Stories They Make Possible: Cognition, Culture, Narrative.* Baltimore: Johns Hopkins University Press, 2008.

INDEX

Abbey, Edward, 83–84, 87, 162, 182
Abortion, 9
Accountability, 17, 18, 90, 93, 178; ecosickness fiction and, 223
Action: environmental, 27, 75, 183, 188, 201–2, 225; ethics and, 12–19, 30, 112, 114; narrative affect and, 7–8, 16, 89, 205–6, 214, 223; revolutionary, 177, 204, 207, 209–10, 213–14; *see also* Activism; Politics
Activism, 4–5, 118, 181–82, 207, 216, 225–26, 228; discord in, 72–76; grassroots, 218–19; *see also* Action; Politics
Activist polemic, 5
Adamson, Joni, 177, 186, 201, 218, 258*n*11, 263*n*91, 264*n*108, 265*n*131
Adaptation, ethic of, 52–54, 75
Addiction, 29, 125, 154–55, 159, 176
Advertising, 1, 4, 117–21, 252*n*38
Aesthetics, 17, 25, 30, 38, 50–54, 64–65, 69, 74, 75, 86–88, 92, 118, 122, 124, 134, 154–55, 164, 222, 228; nature and, 45, 51; thanatological, 51, 53
Affect, 4, 23–26, 30, 219, 225–26, 234*n*65, 240*n*66, 254*n*84; administration of,
26–27; agglomerative, 28; Ahmed on, 25; Altieri on, 25; Berlant on, 25; body-based feelings and, 3; character relations and, 3; connectedness and, 28; contradictions of, 22–23; environmental consciousness and, 2, 22, 225, 233*n*59, 233*n*61, 233*n*63, 267*n*20; environmentalism and, 22; epistemic disposition and, 71; functions of, 15–16, 221, 223, 225; information and, 8; metaphor and, 3; narrative strategies and, 12–13; rhetoric and, 25; sickness and, 15–16, 222; *see also* Emotion; Feelings; Narrative affect; *specific affects*
Affective itineraries, 145–52
Affective turn, 16–17, 23
Affect theory, 16–17, 23–24, 26, 29, 221, 266*n*12
Agency, 13, 17–19, 29–30, 39, 65, 67, 118, 172, 193, 208–15, 222; suspension of, 160, 170, 207, 223; technologization and, 13, 112, 175, 209
Agribusiness, 99
Ahmed, Sara, 24; on affect, 25, 244*n*28; on normativity, 49

AIDS, 4, 34, 56; first cases of, 31; geography of, 39–40; health movement, 39, 73, 75; as imported, 35; mobility of, 31–33; model of, 40; morality and, 40–41; naturalness of, 41; spaces of, 32–34, 37, 56, 64–65; technoscience and, 31; transmission of, 41–42; *see also* People with AIDS

AIDS activism, 72–76; *see also* People with AIDS

AIDS health movement, 72–74

AIDS in Rural America (Briggs), 33

AIDS memoirs: discord in, 25, 27, 64–65, 69–70; nature and, 27–28, 46; nonurban, 27, 35, 38, 65–66, 71, 75–76, 235n6; stance of, 71; *see also* Memoir; People with AIDS

Alaimo, Stacy, 5, 18, 22, 218; *Bodily Natures*, 21, 135, 229n12; on memoir, 55, 68, 242n85

Alcoholism, 9

Allen, Paula Gunn, 178–79

Almanac of the Dead (Silko), 10, 13, 16, 20, 63; affective range of, 30, 170; agency in, 208–15; anxiety in, 29, 30, 116, 160, 167, 168, 174, 177, 204–8, 215; apocalypticism in, 195–204; environmentalism and, 182–83, 184–85, 188; eugenics in, 186–89, 260n51; hyperbole in, 206–7; jeremiad in, 201–4; medicalizing tropes in, 169, 197, 199–200, 204, 222; merging in, 63, 169, 175, 178–80, 182–84, 186; revolution in, 20, 176–77, 179–80, 197, 204, 207, 209–14, 264n110, 264n123; sickness in, 168–70, 175–77, 179, 184–85, 207, 210–11, 213; technology in, 169–70, 174, 185–98, 202, 208–9, 261n57

Altieri, Charles: on aesthetics, 25; on affect, 25, 234n65; on emotion, 66, 234n65; on involvedness, 24; *The Particulars of Rapture*, 228

Altruism, neuroscience of, 101

Anatomy, falsehoods about, 41

Anhedonia, 126, 140–42; *see also* Depression

Animal's People (Sinha), 218

Anthropocentrism, 182–83, 237n29

Anti-interventionism, 8

Antinaturalism, 42–43

Anxiety, 7, 16, 21–22, 226, 263n101; in *Almanac of the Dead*, 29, 30, 116, 160, 167, 168, 174, 177, 204–8, 215; function of, 22–23; Jameson on, 173, 257n5; tone of, 169–70, 173–74, 204–5; as trigger, 30; utopia and, 173, 257n5

Apocalypticism, 19, 30, 196–97, 201–2, 207–8; rhetoric of, 201–4

Appalachian exploitation, 36–37

Art, 25, 87, 120; CAE, 225–26; performance, 225–26; of Wojnarowicz, 55–56, *57–58*

Ashley, Christopher, 33

Attachment, disgust and, 26

Atwood, Margaret, 168, 194, 218

Bacigalupi, Paolo, 194

Balance, 43–44; of detachment, 108, 130, 153; justice and, 130; language and, 50–51; of mind, 80, 106–7; of natural order, 37; ratio, 92; return to, 181; of sickness and health, 179; of skepticism and science, 73; of system, 45; technology and, 172; Wallace and, 153, 163–64; *see also* Harmony; Homeostasis

Barad, Karen, 18

Barkun, Michael, 196

Barth, John, 20

Barthelme, Donald, 20

Beauty, 27, 39, 45, 49–50, 54, 64; *see also* Aesthetics

Beck, Ulrich, 231n31, 231n41

Bell, Michael, 164

Benefit: risk weighed against, 174, 226; of synthetic organisms, 225–26

INDEX   297

Bennett, Jane, 18
Berlant, Lauren, 24, 239*n*42, 266*n*12; on affect, 25, 234*n*67
Bildung, 90, 241*n*69
Biological citizenship, 7, 64, 110–11; varieties of, 22
Biomedical ethics, 6, 67, 102
Biomedicalization, 9, 171, 190–91, 193, 222, 230*n*19, 231*n*23; *see also* Medicalization
Biomedicine, 17, 23–24, 26, 73, 135, 170; obstacles in, 31; PWAs and, 73; *see also* Biotechnology; Technology; Technoscience
Biophilia: The Human Bond with Other Species (Wilson), 22
Biopower, 193
Biosphere, 2, 10, 176, 187–88
Biotechnology, 10, 19, 225–26, 230*n*21, 231*n*23; in *Almanac of the Dead*, 167, 177, 185–86, 191–95, 202, 206, 208–9, 215–16; in *Woman on the Edge of Time*, 29, 172–75, 185, 215; *see also* Biomedicine; Genetic modification; Technology; Technoscience
Biovalue, 193
Bloch, Ernst, 204–5
Bodies, 2–3, 9; ecological view of, 135; environment and, 5, 7, 10, 45, 46, 74, 134–35, 138–39, 143; medicalization of, 6; as metonym, 45; norms of, 41–43, 45, 74; tropes of, 30, 48–49, 50, 140, 143–44, 199–200; in *Woman on the Edge of Time*, 185–86
Bodily Natures (Alaimo), 21, 135, 229*n*12
Body-based feelings, 3, 17, 38, 39, 67; *see also* Feelings
Body-environment boundary, 3, 5, 24, 52–53, 60, 123–24, 222
Body-land merging, 63, 169, 175, 178–80, 182–84, 186; *see also* Merging narrative
Bog ecosystems, 51

Botkin, Daniel, 43–45, 51, 238*n*32
Bourdieu, Pierre, 161
Bourjaily, Vance, *Old Soldier*, 33
Brain science, *see* Neuroscience
Brennan, Teresa, 254*n*84
Briggs, Richard, *AIDS in Rural America*, 33
Buell, Frederick, 207
Buell, Lawrence, 21–22, 118, 151, 202, 204, 232*n*42, 233*n*61, 246*n*48; *The Environmental Imagination*, 90–91; on *flânerie*, 146, 147, 151; on toxic discourse, 139; *Writing for an Endangered World*, 21

CAE, *see* Critical Art Ensemble
Cameron, Sharon, 100
Capgras syndrome, 80–84; love in, 105; paranoia and, 107–8; ramifications of, 102–4; symptoms of, 94–95, 97
Capitalism, 29, 194, 203–4, 210, 219; neoliberal, 175; optimism and, 220–21
Carson, Rachel, 79, 87, 145, 162, 201; on environmental experience, 84; environmental legislation and, 8; *The Sense of Wonder*, 83; *Silent Spring*, 5, 8, 22, 135, 137
Causality, 2–3, 6, 17, 20, 100, 124, 145, 163; wonder and, 80
Ceremony (Silko), 167, 176, 179, 199, 216, 218, 259*n*23
Chakrabarty, Dipesh, 230*n*16
Character relations, affect and, 3
Chemical exposure, 1; *see also* Multiple chemical sensitivity
Chemical production, long-term effects of, 1–2
Cheng, Anne, 69, 70, 75
Clean Air Act (1963), 8
Clean Water Act (1977), 8
Climate change, 8, 117–18, *119*, 217–18, 220, 225, 232*n*52, 267*n*21
Climate summit, 218

298 INDEX

Close to the Knives: A Memoir of Disintegration (Wojnarowicz), 27, 33, 55–70, 72, 222, 224, 239n52; conceptual boundaries in, 74–75; nature in, 43, 46, 60–61, 63–64; sickness in, 38–39, 59, 64
Cognition, 3, 28, 80, 87, 93, 97–98, 115, 116; cultural, 227
Cognitive map, 141
Colonialism, 194, 198; censure of, 203; neocolonialism, 181, 186, 204; overthrowing, 29, 178; slavery and, 175; *see also* Imperialism
Commoner, Barry, 105, 201
Communitarianism, 227
Computerization, 9
Connectedness, 29, 81; affect and, 28; in *The Echo Maker*, 80–81, 85, 101, 105–8, 114–15; ecological model of, 12, 80, 105; intimacy of, 14–15; Powers on, 28; Wallace on, 28, 163; *see also* Body-land merging; Human-environment imbrication; Interconnectedness
Conservation, 79, 88; *see also* Preservation
Consumerism, 122, 133, 137–38; *see also* Consumption
Consumption, 115, 130, 137, 161, 166; prodigal, 131–34; renouncing, 134; *see also* Consumerism
Contagion, 103, 107, 153
Contagious: Cultures, Carriers, and the Outbreak Narrative (Wald), 20
Contamination, 6, 137, 139–40; quarantine, 133–35; threat of, 186–87; *see also* Pollution
Contested natures, 39–46; *see also* Nature
Correlation, 100, 222
Cousins, Mark, 160
Cranes, *see* Sandhill cranes
Critical Art Ensemble (CAE), 225–26
Cronon, William, 47, 52

Cultural Cognition Project, 225–27
Cultural production, 6, 13
Cunningham, Michael, 33
Cvetkovich, Ann, 23, 266n15; *Depression*, 221; on *North Enough: AIDS and Other Clear-Cuts*, 47; on secondary traumatic stress, 46
Cyborg manifesto, 111–12

Damasio, Antonio, 234n65
Daston, Lorraine, 77
de Certeau, Michel, 151
de Crèvecoeur, J. Hector St. John, 99
Deep ecology, 42, 134, 189–90; *see also* Ecocentrism
Defamiliarization, 80, 87, 91–92, 100; disgust and, 28, 124, 159, 163; modernist, 88–90; wonder and, 93, 115
DeLillo, Don, 6
Depression, 12, 221; agony of, 129; detachment and, 140; in *Infinite Jest*, 140–42; sources of, 129; species extinction and, 4; *see also* Anhedonia
Depression (Cvetkovich), 221
Descartes, René, 78, 89
Description, 36, 104–5, 111, 143–45, 192, 199–200, 222; disgust in, 138–39, 155–56; of disorder, 94; focalization of, 98–99
Desert, 63–64, 67, 194, 198–201, 262n71
Detachment, 17–18, 124, 126–27; balance of, 108, 130, 153; depression and, 140; in *Infinite Jest*, 29, 108, 128–31; risk of, 125
Determinism: biological, 208–9; geographical, 98–99
Di Chiro, Giovanna, 74
Dillard, Annie, 178
Dimock, Wai Chee, 7, 267n19
Discord, 38; in activism, 72–76; in AIDS memoirs, 25, 27, 64–65, 69–70; destabilization of, 39, 67–68; effect

of, 68–69; ethics and, 75; feelings of, 65–72; inspiration through, 25; intensity of, 66–67; model of, 69–70; in *North Enough: AIDS and Other Clear-Cuts*, 65–66, 72; as psychic disturbance, 72; schematics of, 71–72; sickness and, 65–66; suspicion and, 27, 39, 68–69, 71–72; workings of, 66

Disease, 11, 27, 31, 45, 60, 63–64, 174, 222, 228; as analytic, 20–21; environment and, 38–39; *see also* Illness; Sickness; *specific diseases*

Disgust, 18, 22, 27, 108, 118, 120, 204; attachment and, 7, 26, 116, 223; defamiliarization and, 28; in descriptions, 120, 155–56; embodiment of, 158–59; in environmentalism, 117–22, 225; in *Infinite Jest*, 29, 116, 124, 137, 152–65, 255n100

Displacement, patterns of, 36

Earth: hope for, 215; perception of, 27; soma and, 2–3, 27, 65, 74, 143–44, 222

Earth Day, 8; 1990, 14

The Echo Maker (Powers), 11, 16–17, 28, 76, 79–90, 100–1, 111–16; care in, 80–81, 89, 94, 100–1, 104, 107–8, 110, 112, 114–15; connectedness in, 81, 104, 106–8, 116; defamiliarization in, 80, 87, 88–90, 91–92, 93, 100, 115; environment and, 78–79, 83, 86, 89, 93, 97–99, 103, 105, 113; as nature writing, 80, 83, 100; paranoia in, 81, 101, 107–8, 115–16, 152, 165; plot of, 93–94, 97, 163; projection in, 28, 50, 81, 101, 107–8, 114–15; sickness in, 79–80, 95, 114; wonder in, 28, 76, 79–81, 83, 88–95, 98, 100, 108, 114–15, 163

Ecocentrism, 183; *see also* Deep ecology

Ecocriticism, 8, 21–23, 30, 165, 168, 195, 216, 224–25, 237n31; responsiveness of, 26–27; *see also* Environmental criticism

Ecological endangerment, 3–4, 7, 13, 39, 222; *see also* Endangerment; Environmental endangerment

Ecological illness, somatic illness and, 222–23

Ecological perspective, the, 106, 108; central tenet of, 104–5; detection of, 135–36; elaboration of, 14–15

Ecosickness fiction, 2–3, 10–11, 14–19, 48–49; affect in, 7, 26, 39; belonging in, 20; as literature of accountability, 223; post-1970s, 5; production of, 19–20; sickness in, 12; technologization and, 13; *see also* Fiction; Science fiction

Egalitarianism, 227

Ehrenreich, Barbara, on optimism, 220–21

Ehrlich, Paul, 79, 201

Eiseley, Loren, 178, 246n50

Emotion, 15, 23, 116, 234n65; Altieri on, 66; attachment and, 24; in ecosickness fiction, 7; ethicopolitical effects of, 13, 221; literature and, 27; Ngai on, 24, 26–27; risk and, 226–27; Thornber on, 25–26; volatility of, 25–26; *see also* Affect; Feelings; *specific emotions*

Empiricism, 4, 6–7; literature and, 15, 20, 30, 227

Endangerment, 3, 12, 17, 26–27; ethical responses to, 4; forms of, 4; toxic, 2; *see also* Ecological endangerment; Environmental endangerment

The End of Nature (McKibben), 14, 22

Environment, 33; built, 96–97, 140, 150–51; disease and, 38–39; embodied engagement with, 4; harmony of, 37–38; risk and, 1, 10; *see also* Lived environment; Nature

Environmental awareness, 28, 78; *see also* Environmental consciousness

Environmental consciousness, 19, 223; affect and, 2, 22, 93, 116, 124; medicalized, 38–39, 93, 124, 134–35, 145, 153, 197; *see also* Environmental awareness
Environmental criticism, 16; *see also* Ecocriticism
Environmental damage, 16; urgency of, 113; *see also* Environmental degradation; Environmental endangerment
Environmental degradation: technology and, 8–9; *see also* Environmental damage; Environmental endangerment
Environmental discourse, 29–30, 78, 108, 178, 202, 218–19
Environmental endangerment, 39; agency and, 13; conditions of, 13, 21; *see also* Environmental damage; Environmental degradation
Environmental health movement 39, 74–75; *see also* AIDS, health movement
Environmental history, 20–21
Environmental illness, WHO on, 1
Environmental imagination, 5–6, 13, 35, 78, 177, 179; contemporary, 78, 202; medicalization of, 29; of toxification, 2, 145, 147; *see also* Buell, Lawrence
Environmentalism, 8–10, 14–15, 93, 134, 165, 178, 207, 219; affect and, 22; *Almanac of the Dead* and, 182–83, 184–85, 188; praxis of, 225; racism of, 189–90
Environmental justice movement, 39, 72, 178, 190, 216, 241*n*84
Environmental policy, 5, 44
Environmental toxicity, 2, 135, 139; memoirs of, 6, 229*n*12; *see also* Toxification
Environmental writing: canon of, 84; expectations for, 76; health in, 238*n*33; identification in, 178–79; invention in, 91; merging narrative in, 63, 169, 178; mimesis in, 91; perception and, 83–84; tropes in, 221–22; *see also* Nature writing
Epidemiological stories, 20
Epstein, Julia, 11
Epstein, Steven, 73
Estok, Simon, 233*n*61, 267*n*20
Estrangement, 100
Ethic of care, 28, 54–55, 83; solipsism and, 109
Ethics, 6, 8, 43, 53, 68, 75, 162, 219, 223, 227; action and, 12–19, 30; of as-is, 54, 75, 224; of place, 10
Ethnic purification, 189–90; *see also* Eugenics
Etiology, 2, 6
Eugenics: banner of, 186; origins of, 186–87; preservation and, 186–87; research, 194–95; taint of, 208; *see also* Ethnic purification
Evaluative templates, 24–25
Expansionist hubris, 22
Expository mode, 89

Fear, 186, 201, 205, 267*n*20; biotechnological risk and, 226
Feelings, 6, 150–51, 217, 234*n*65; discordant, 65–72; environmental, 22, 199; *see also* Affect; Body-based feelings; Emotion
Felski, Rita, 69–70, 75, 241*n*76
Feminism, 21, 42
Fetterley, Judith, 49
Fiction, 3, 20–21, 23, 124, 126–27, 222; as laboratory, 19; making, 191; mobilizing power of, 16; speculative, 29, 168, 175, 191; *see also* Ecosickness fiction; Science fiction
Fisher, Philip, 66, 91, 236*n*11, 245*n*37
Flâneur, 146–47; *see also* Walking
Form, 11; of agency, 193; apocalyptic, 205, 213–14; of authority, 72–73; of care, 54, 184; of creativity, 25; of

power, 193; pressure on, 17; problems of, 4–5; of sickness, 169; of suspicion, 27, 39; tropes and, 128
Foucault, Michel, 43; *The History of Sexuality*, 42
Freedom, 112, 127, 164; illusion of, 29
Free market fundamentalism, Hawken on, 218
Fuller, Robert, 89

Gain (Powers), 6, 79, 101; plot of, 2–3
Garrard, Greg, 238*n*33
Gender, 1, 12, 145, 168, 170, 173, 193, 231*n*31
Gene splicing, 9–10, 173, 226, 230*n*17
Gene therapy, 7, 9, 167, 191
Genetic modification (GMOs), 4, 9–10, 44, 173, 185, 226–27; debate about, 9, 174; *see also* Genetics
Genetics, 14, 80, 109; as code, 4, 81, 82; manipulation of, 4, 187; in *Woman on the Edge of Time*, 169, 173, 191, 194–95; *see also* Gene therapy; Genetic modification
GenTerra project, 225–27; *see also* Critical Art Ensemble
GMOs, *see* genetic modification
The Gold Bug Variations (Powers), 28, 79–82, 86–87, 90, 93, 108–9, 112
Gould, Stephen Jay, 40–41, 42
Grassroots activism, 74, 218–19
Grosz, Elizabeth, 140, 143
Grover, Jan Zita, 4, 7, 20, 25–26, 49, 67, 107, 178, 228, 239*n*44, 242*n*85; as caregiver, 46; environmental vision of, 47–48, 143; on landfill, 52–53; on nature, 40, 53–54, 75, 219; *North Enough: AIDS and Other Clear-Cuts*, 18, 27, 33, 34, 38–39, 43, 46–55, 64–66, 68, 70–71, 74–76, 200–1; PWAs and, 48, 54–55; on sickness, 50–51; suspicion for, 69; thanatological aesthetic of, 51, 53; voice of, 57; Wojnarowicz compared with, 63–65

Guha, Ramachandra, 190

Habit, banality of, 217
Hacking, Ian, 102, 194–95
Haidt, Jonathan, 120
Halberstam, Judith, 33
Hale, Dorothy, 81
Haraway, Donna, 111; on nature, 14
Harmony, 15, 197, 224; discordance and, 43–44; of environment, 37–38, 67, 69, 106–7; ideals of, 44, 65; metric of, 44; nature and, 27, 45, 61; nostalgia and, 67; somatic, 237*n*32; telos of, 44; *see also* Balance; Homeostasis
Hawken, Paul: on free market fundamentalism, 218; on hope, 218–19
Hayles, N. Katherine, 131, 135–36
Haynes, Todd, 2, 6
Health, 27, 35, 65, 189, 227; in environmental writing, 238*n*33; nature and, 42, 44–46, 49, 53–54, 63; sickness balanced with, 179; types of, 73–74; as unnatural, 237*n*32
Heise, Ursula K., 10, 178, 195, 233*n*61
Hermeneutics of suspicion, 26, 69–70; *see also* Suspicion
Herring, Scott, 33
Hierarchism, 227
Hirstein, William, 94
The History of Sexuality (Foucault), 42
HIV, *see* AIDS
A Home at the End of the World (Cunningham), 33
Homeostasis: veracity of, 106; *see also* Balance; Harmony
Homology, 3, 45, 222
Homosexual identity formation, 33
Homosexuality, 9; as metrocentric, 32–33; nature and, 40–42
Hope, 26, 54, 67, 109, 114, 118, 133, 155, 165, 208–9; endorsement of, 219; environmental, 215, 218; fraud of, 203–4; Hawken on, 218–19; for justice, 175;

Hope (*continued*)
  loss of, 218; misguided, 131; yearning for, 221; *see also* Optimism
Human-environment imbrication, 5, 11, 13–14, 18, 21, 52, 59, 80, 89, 139, 199, 201
Human exceptionalism, 18
Human Genome Project, 14
Humans, privileging, 14
Hyperbole, 153, 162–63, 206–7

Iatrogenesis, 104, 248n68
Identity, 37, 42, 146, 151, 186, 212–13; of ecocriticism, 224–25; homosexual, 33; illness and, 11; protection of, 227; sexual, 41
Illich, Ivan, 230n19, 237n32, 248n68
Illness, 2, 10–11, 249n87; embodied experience of, 6; Sontag on, 111; *see also* Disease; Sickness; Somatic illness
Imperialism: forms of, 125; gains of, 220; meeting demands under, 131–32; slavery and, 177; *see also* Colonialism
Individualism, 209; critique of, 114; isolated, 133; liberal, 114; as preference category, 227
Industrialization, long-term effects of, 1–2, 138
Industrial mining, 5
Industrial pollutants, 74, 138; *see also* Pollution; Toxicity; Toxification
Infinite Jest (Wallace), 11–12, 20, 26, 110, 120–26, 200, 222, 223; connection in, 28, 122, 124–25, 127, 130, 135–36, 146, 149; depression in, 129, 140–42; description in, 138–39, 143–46, 155–56; detachment in, 29, 108, 125–31, 146; disgust in, 29, 116, 120, 124, 137, 152–65, 245n37, 255n100; energy production in, 15, 131, 133, 137–38, 166; environment in, 122, 123–24, 130–39, 142, 145, 151, 155, 161–62, 164–65; experialism, 138–39, 166; hyperbole in, 153, 162–63; interdependence in, 131, 142, 148–50, 152, 163; medicalization of space in, 140, 143–45, 153, 164; passive voice in, 125, 128; sickness in, 124, 134, 139–40, 153, 157; solipsism in, 29, 124–27, 146–47, 151, 153, 164, 255n100; walking in, 146, 150–52
Information, affect and, 8, 217
Innovation, literary, 11, 17, 47, 68, 91, 163–64
Interconnectedness, 28; connotations of, 15; risk and, 15; systems and, 80–81; *see also* Body-land merging; Connectedness; Human-environment imbrication
Involvedness, Altieri on, 24

Jameson, Fredric, 130, 141–43, 173, 257n5
Johnson, E. Patrick, 56
Journalism, *see* Popular journalism
The Jungle (Sinclair), 5
Justice, 189–90, 211, 214, 218; technoscience and, 169, 175, 224; *see also* Environmental justice movement

Kakutani, Michiko, 154
Kenyon College commencement speech (2005), 164
Killingsworth, M. Jimmie, 201–2
Kingsolver, Barbara, 218
Kleinman, Arthur, on sickness, 11
Kolbert, Elizabeth, 218
Kolodny, Annette, 145
Krauss, Rosalind, 251n20
Kristeva, Julia, 156–57, 160, 161

Laissez-faire ethics, 8
Land use, 60, 75, 81, 83, 97, 113, 168, 190–91, 198–199
Langone, John, 41, 236n18
Langston, Nancy, 20–21, 232n56
Language, balance and, 50–51
Latour, Bruno, 14, 69, 241n75

INDEX   303

Leopold, Aldo, 47
Le Sueur, Meridel, 5
Life itself: conception of, 9, 13–14, 30, 208; history of phrase, 4; intervention in, 5, 10, 14, 173, 176, 185, 195, 206, 215–16, 224, 225–26; malleability of, 14, 193, 207; reimagination of, 4, 7, 13–14
Liminal Lives (Squier), 19–20
Liminal zones, 61–62
Linnaeus, Carl, 78
Literature, 4–7; emotion and, 27; empiricism and, 15, 20, 30, 227; *see also* Ecosickness fiction; Fiction; Innovation, literary; Memoir; Narratives; Science fiction
Lived environment, 27, 54, 222; *see also* Environment; Place; Space
Localism, 10, 178
Lyrical mode, 28, 89–90, 92–93, 113

Macrosocial forces, 11–12, 166
Marsh, George Perkins, 44
Marx, Leo, 15, 99
Massumi, Brian, 201, 234n65, 240n66
Master tropes, 40, 44; *see also* Tropes
Max, D. T., 154
McCauley, Clark R., 120
McKibben, Bill, 14; *The End of Nature*, 22
MCS, *see* Multiple chemical sensitivity
Medical industrial complex, 9
Medicalization, 7, 9, 144, 230n19, 248n68; of bodies, 6, 138–40, 162, 169; of environmental consciousness, 29, 38–39, 93, 124, 134–35, 144–45, 153, 197; of space, 17, 140, 143–45, 151, 153, 164, 169, 197, 199–200, 204, 222; *see also* Biomedicalization
Medicine, 2, 9, 31–32, 35, 45, 53, 73–74, 112, 135, 171–73, 185, 192, 235n3, 249n87; *see also* Biomedicine; Medicalization; Regenerative medicine

Memoir, 19–20; Alaimo on, 55, 68; of environmental toxicity, 6, 229n12; as genre, 39, 47, 68, 72, 76; of immigration, 34, 38; language in, 50–51; *see also* AIDS memoirs
Merging narrative, 62; in environmental writing, 63, 178; *see also* Body-land merging; Narratives
Metafiction, 79, 126, 244n30, 255n97
Metaphor, 15, 17, 62, 91–92, 105–7, 111, 142, 144, 182, 184, 197–201, 222; affect and, 3; in *North Enough: AIDS and Other Clear-Cuts*, 200–1; suspicion of, 69; tone of, 106–7; *see also* Medicalization
Metonymy, 32, 62, 171, 180–81; body as, 45; criticism as, 70; *see also* Synecdoche
Metronormativity, 33
Michaels, Walter Benn, 212–13
Midwest, the, 97, 99–100
Miller, D. A., 254n88
Miller, William Ian, 159, 161, 255n101
Mimesis, in environmental writing, 91, 162; in *Infinite Jest*, 162–63
Misery lit, 47, 224, 239n52
Mitman, Gregg, 20–21, 231n28, 232n56
Miyoshi, Masao, 267n19
Mobility: of AIDS, 31–33, 235n5; class, 114, 203; personal, 59–60; of Wojnarowicz, 59–60
Modernization, 21, 92, 113; defamiliarization and, 88–90; metropolitan, 147–48; reversing, 134, 196; technique, 87–88; technological, 61
Molecular biology, 4
Morality: AIDS and, 40–43, 63; disgust and, 120, 159, 161; nature and, 27, 40–42
More-than-human, 3, 10, 12, 18, 21–24, 30, 36–37, 43, 68, 93, 210, 225; human and, 45, 61, 74–75, 80, 106, 108, 116, 147, 152, 161, 169, 179; value of, 71; *see also* Environment; Nature; Nonhuman

Mortimer-Sandilands, Catriona, 54
Morton, Timothy, 72, 120
Muir, John, 63, 83, 248*n*71
Multiple chemical sensitivity (MCS), 1, 4, 6
Mumford, Lewis, on suburbs, 60
Muñoz, José, 221, 266*n*15
My Own Country: A Doctor's Story (Verghese), 32–39, 67, 241*n*69

Narrative affect, 2–4, 19, 29, 228; action and, 16; ecological systems revealed by, 30; ethics and, 7, 15–16, 81; sickness and, 22, 26, 216, 222; *see also* Affect
Narrative illiteracy, 7
Narratives, 3–4, 7, 11, 20, 79, 112, 176; affect and, 12–13; environmental, 30, 83, 113, 139; etiology and, 2; of outbreak, 20; of progress, 13, 31, 185, 195, 196, 202, 215, 218; *see also* Fiction; Literature; Memoir; Merging narrative
Nash, Linda Lorraine, 20–21, 135, 232*n*56
National Oceanic and Atmospheric Association (NOAA), 217–18
Nature, 13, 49, 51, 65; aesthetics and, 45, 51; AIDS memoirs and, 27–28, 46; beauty and, 39, 46, 49–50, 64; body and, 46; in *Close to the Knives: A Memoir of Disintegration*, 43, 46, 60–61, 63–64; as concept, 39, 43, 46, 54, 63, 72, 75; without ecology, 72; Grover on, 40, 53–54, 75, 219; Haraway on, 14; harmony and, 27, 45, 61; health and, 42, 44–46, 49–50, 53–54, 63, 64; morality and, 27, 40–42; normalized meaning of, 27, 34, 43–44, 48, 65, 67, 68, 75; in *North Enough: AIDS and Other Clear-Cuts*, 43, 48–54; order in, 37, 40, 42, 44; Wojnarowicz on, 60–62, 219; *see also* Contested natures; Environment

Nature writing, 5, 14–15, 17, 19, 61, 83, 100; hermetic, 47–48; Slovic, S., on, 116, 231*n*38; wonder in, 28, 78; *see also* Environmental writing
Neoliberalism, management under, 167; *see also* Capitalism, neoliberal
Neurological realism, 83, 243*n*13
Neuroscience, 100, 234*n*65, 245*n*32; of altruism, 101; neo-Cartesian, 102; of optimism, 220; research, 90, 101
Newton, Isaac, 77–78, 80, 86
Ngai, Sianne, 3, 160, 205, 207, 215, 234*n*65; on emotion, 24, 26–27; *Ugly Feelings*, 26
Nietzsche, Friedrich, 73
Nihilism, 8, 221
Nixon, Rob, 13; *Slow Violence and the Environmentalism of the Poor*, 21–22, 233*n*60
NOAA, *see* National Oceanic and Atmospheric Association
Nonhuman, 48–49; human and, 14, 18, 61, 65, 88, 158, 222; *see also* More-than-human
Norms, 25, 69; of nature, 27, 34, 43–44, 48, 65, 67, 68, 75
North Enough: AIDS and Other Clear-Cuts (Grover), 18, 27, 33, 46, 48–55; beauty in, 45, 49–50, 54; conceptual boundaries in, 74–76; Cvetkovich on, 47; discord in, 65–66, 72; ethic of the as-is, 54, 75, 224; metaphor in, 143, 200–1; Mortimer-Sandilands on, 54; nature in, 43, 48–54, 74; sickness in, 38–39; suspicion in, 39, 70–71; thanatological aesthetic in, 51, 53; tone in, 64–65; voices in, 68
Nostalgia, 53, 63, 67, 120
Novels, 213–14; functions of, 28, 122, 123, 164, 165; Powers on, 244*n*30; Wallace and, 126–27, 154, 158; *see also* Ecosickness fiction; Fiction; Narratives; Science fiction

Nuclear: energy, 10, 227; weaponry, 167, 196, 257*n*1
Numbness, 141, 163–64

Old Soldier (Bourjaily), 33
Ontology, 18, 151, 206
Optimism, 227, 266*n*15; bias, 220; capitalism and, 220–21; cruel, 25–26, 234*n*67, 266*n*12; dangers of, 219–20; *Ecosickness* and, 219, 223; Ehrenreich on, 220–21; Hawken on, 218–19; range of thought on, 221–22; techno-optimism, 13, 185, 209, 216, 219; *see also* Hope
Optimization, 9, 110, 208; consequences of, 10

Packard, Vance, 130
Palmer, Jacqueline, 201–2
Palumbo-Liu, David, 152, 234*n*68, 254*n*84
Paranoia, 70, 71, 81, 152; Capgras syndrome and, 107–8; in *The Echo Maker*, 81, 101, 107–8, 115–16, 152, 165
Park, Katharine, 77
Parody, 186–87, 226; dark, 10, 188
The Particulars of Rapture (Altieri), 228
Passivity, 205, 207–8
Patient, as client or consumer, 9–10; *see also* People with AIDS
PCBs, *see* Polychlorinated biphenyls
People with AIDS (PWAs): on biomedicine, 73; Grover and, 47–48, 54–55; treatment of, 35; Verghese and, 31–32; *see also* AIDS; AIDS activism; AIDS memoirs
Perception, 16, 19, 29, 81, 87, 89, 115; defamiliarization and, 87–88; environmental, 3, 34, 38, 51, 65, 94–96, 199, 222; environmental writing and, 83–84; sickness and, 11, 38, 48, 66, 71, 91; transformation of, 23, 60, 83–85, 116

Pessimism, 218; technopessimism, 195; wonder and, 85, 116
Phenomenology, 23
Phillips, Dana, 44–45, 106–7, 237*n*31
Piercy, Marge, 4, 7, 165; *Woman on the Edge of Time*, 20, 29, 168–75, 185–86, 194–95, 200, 207, 214–15, 221
Place, 33–34, 80, 94–98, 246*n*48; -based perceptual ecology, 100; -bashing, 199, 262*n*73; ethics of, 10, 178; non-places, 97; *see also* Environment; Space
Planetary consciousness, 8; *see also* Environmental consciousness
Pole, David, 120, 161
Politics, 11, 43, 213, 221, 224; affect and, 3, 6, 13, 15, 23–27, 39, 72, 160, 221, 234*n*65, 241*n*76; environment and, 27, 30, 57, 88, 131, 134, 137–38, 142, 180; medical, 73; *see also* Action; Ethics
Pollution, 5, 36, 74, 134, 252*n*35; racism and, 190, 260*n*43; rhetoric of, 189–90; *see also* Contamination; Industrial pollutants; Toxification
Polychlorinated biphenyls (PCBs), 12
Poor, the, 22, 74, 168–69, 173, 175, 186–87, 190–91, 218
Popular journalism, 4
Positivity, dangers of, 221
Posthumanism, 8, 18–19, 232*n*47–48
Postmodernism, 14, 17, 19–20, 79, 88, 130, 164; criticism of, 124–25, 127; irony, 136; techniques of, 4–5, 158, 244*n*30, 246*n*47
Poststructuralism, 42, 46, 79
Powers, Richard, 4, 6–7, 20, 76, 92, 100–1, 109, 124, 220, 223, 228, 243*n*9, 243*n*13, 244*n*30, 245*n*37, 245*n*40, 250*n*93; on connectedness, 28, 105; *The Echo Maker*, 11, 16–17, 28, 76, 79–90, 108, 112, 116; on fiction, 101, 223, 244*n*30; on focalization, 90, 97; *Gain*, 2–3, 6, 79, 101; *Galatea 2.2*,

Powers, Richard (*continued*)
101, 245*n*40; *The Gold Bug Variations*, 28, 79–82, 86–87, 90, 93, 108–9, 112; writing patterns of, 81
Preservation, 8, 190; eugenics and, 186–87; water, 115; *see also* Conservation; Environmentalism
Prodigal Summer (Kingsolver), 218
Profit, 82, 191–92, 252*n*29
Progress, *see* Narratives, of progress
Projection, 28, 50, 81, 101, 107–8, 114–15
Prophecy, 178, 180, 201–2, 211, 214, 216, 261*n*57, 263*n*91
Pryse, Marjorie, 49
Psychoanalytic theory, 23
PWAs, *see* People with AIDS
Pynchon, Thomas, 20, 127

Queer activism, 42, 74
Queer fiction, 49–50
Queer theory, 20, 23, 33, 49, 56, 221, 236*n*22, 266*n*15

Rabinow, Paul, 14, 193
Racism, pollution and, 190, 260*n*43
Ramachandran, V. S., 94
Realism, 5, 79, 83, 120, 158, 225, 237*n*31, 243*n*13, 244*n*30, 246*n*47, 255*n*97
Reciprocity, 88; transpersonal, 146–47, 149
Recombinant DNA, 9, 10, 19, 173, 226, 230*n*17, 230*n*21; *see also* Genetic modification
Reed, T. V., 210, 258*n*11, 258*n*19
Regenerative medicine, 9, 185, 191–92, 195; *see also* Gene therapy
Regionalism, 49, 267*n*19
Reparative reading, 26, 75
Representational conventions, 29, 33, 64, 74, 199, 225
Representational dilemmas, 4–5, 19, 26
Repulsion, 152, 155, 199; *see also* Disgust

Responsibility, 18–19, 184, 185, 221
Rhetoric: affect and, 25, 68; of apocalypticism, 201–4; environmental, 39, 118, 120, 169, 178–79, 187, 216; of pollution, 189–90; of Wojnarowicz, 57, 59
Ricoeur, Paul, 69, 163
Risk: assessment, 226; benefit weighed against, 174, 226; emotion and, 226–27; environment and, 1, 6, 10; exploring, 226–27; factors, 35; fate, 231*n*31; fear and, 226; as generative, 233*n*61; individualism and, 227; interconnectedness and, 15; potential, 146; societal, 6, 227; of synthetic organisms, 225–26
Rizzolatti, Giacomo, 103, 104
Rose, Nikolas, 9–10, 110, 185, 193, 208–9
Ross, Andrew, 257*n*2, 260*n*43
Rotello, Gabriel, 41–42
Rozin, Paul, 120
Rural: AIDS and, 34–35; pastoral and, 35–37; -urban dichotomy, 33, 37; Verghese on, 35–37, 67
Ryman, Geoff, 33

Safe (Haynes), 6; plot of, 2–3
Sandhill cranes, 83, 89–90, 97–99, 114; tourism, 97, 113, 115–16
Science, 6–7, 15, 28, 79, 82, 85, 135, 188, 194, 224, 226; balance of skepticism and, 73; purpose of, 86–87, 112; source of, 78, 227; *see also* Biotechnology; Genetic modification; Genetics; Medicine; Technoscience
Science fiction, 19; *see also* Ecosickness fiction; Fiction
Science writing, 4, 5
Scientific Revolution, 78
Sedgwick, Eve Kosofsky, 24, 69–70, 71, 107, 249*n*76; reparative reading, 26, 75
Self-perception, 11
Sentimentalism, 22, 34–38, 45, 50, 63, 106

Serres, Michel, 13
Sexual identity, 41
Sexuality, 42, 56
Shklovsky, Viktor, 87–88
Sickness, 3–4, 6, 12, 17, 30, 169; affect and, 15–16, 222; in *Almanac of the Dead*, 168–70, 175–77, 179, 184–85, 207, 210–11, 213; as analytic, 3, 12, 165, 216, 223–25; in *Close to the Knives: A Memoir of Disintegration*, 38–39, 59, 64; conceptual update of, 11; differing applications of, 12; discord and, 65–66; double nature of, 15; in *The Echo Maker*, 79–80, 95, 114; Grover on, 50–51; health balanced with, 179; in *Infinite Jest*, 124, 134, 139–40, 153, 157; Kleinman on, 11; narrative affect and, 22, 26, 216, 222; in *North Enough: AIDS and Other Clear-Cuts*, 38–39; perception and, 11, 38, 48, 66, 71, 91; perspective offered by, 12; as pervasive, 2, 11, 12, 17, 27, 106, 139, 168–69, 175, 216, 228; space and, 39, 59–60; technoscience and, 8–12; tropes, 3–4, 221; *see also* Disease; Ecosickness fiction; Illness
Silent Spring (Carson), 5, 22, 135, 137; environmental legislation and, 8
Silko, Leslie Marmon, 4, 6–7, 26, 106, 165, 167; *Almanac of the Dead*, 10, 13, 16, 18, 20, 29–30, 63, 116, 160, 167–68, 169–70, 184–85, 194–95, 210; *Ceremony*, 167, 176, 179, 199, 216, 218
Simmel, Georg, 146–47, 152
Simpson, J. H., 199
Sinclair, Upton, 5
Sinha, Indra, 218
Sinigaglia, Corrado, 103, 104
Slavery: colonialism and, 175; imperialism and, 177
Slovic, Paul, 8
Slovic, Scott, 8, 116, 162; on nature writing, 231n38

Slow Violence and the Environmentalism of the Poor (Nixon), 21–22
Smith, Neil, 61, 142, 253n48
Smith, Rachel Greenwald, 79, 244n30, 246n47
Snyder, Gary, 178
Societal risk, 6, 227
Soja, Edward, 142–46
Solipsism, 255n100; critique of, 129–30; ethic of care and, 109; in *Infinite Jest*, 29, 124–27, 146–47, 151, 153, 164, 255n100
Soma: environment and, 2–3, 5, 27, 38, 51, 65, 74, 81, 143–45, 179, 184, 197, 201, 222; *see also* Bodies
Somatic illness, *see* Disease; Illness; Sickness
Somatic vitality, 12
Somatic vulnerability, 3, 147, 162
Sontag, Susan, 39–40; on illness, 111
Soper, Kate, 42–43, 236n10
Southwest, the, 63, 67, 175–76, 198, 200, 260n43
Space, 142–47, 151, 225; of AIDS, 32–34, 37, 56, 64–65; medicalization of, 17, 140, 143–45, 151, 153, 164, 169, 197, 199–200, 204, 222; reconfiguration of, 133, 139, 142, 175, 194; sickness and, 39, 59–60; soma and, 5; toxification of, 131–32, 139; *see also* Lived Environment; Place
Species extinction, 8, 79–80, 85, 222; depression and, 4
Speth, Gus, 223, 225, 227, 267n21
Splicing, gene, *see* Recombinant DNA
Squier, Susan, 19–20
Subject-object distinction, 24
Suspicion, 68–75; discord and, 27, 39, 68–69, 71–72; as epistemic orientation, 27, 70–71; experience preceded by, 71; form of, 27, 39; for Grover, 69; of metaphor, 69; in *North Enough: AIDS and Other Clear-Cuts*, 39,

Suspicion (*continued*)
70–71; *see also* Hermeneutics of suspicion
Sympathy, 149, 151–52
Synecdoche, 129, 212; anatomical, 199; partial, 180; *see also* Metonymy
Synthetic organisms, 225–26; *see also* Genetic modification

Technological intervention, 13–14, 170, 175–76, 195–97, 238*n*32; *see also* Life itself, intervention in
Technological modernization, 61
Technological rationality, 4
Technologization, 5–7, 22, 169, 222; agency and, 13, 112, 175, 209; of bodies, 3–4; ecosickness fiction and, 13; reflexivity of, 15; *see also* Biomedicalization
Technology, 13–14, 147; in *Almanac of the Dead*, 169–70, 174, 179, 185–98, 202, 208–9, 261*n*57; balance and, 172; environmental degradation and, 8–9; perception and, 79; *see also* Biomedicine; Biotechnology; Technologization; Technoscience
Technomorphism, 240*n*62; zoomorphism and, 111
Techno-optimism, 13, 219
Technoscience, 8, 10, 13–15, 17, 20, 26, 109, 112, 167, 169, 173–75, 195–97, 208–9, 216, 221, 230*n*21; AIDS and, 31; inequalities of, 29–30, 173, 186, 188, 191; justice and, 169, 175, 224; sickness and, 8–12; *see also* Biomedicine; Biotechnology; Life itself, intervention in; Science; Technology
Thacker, Eugene, on life itself, 4
Thomashow, Mitchell, 83–84, 87, 100, 162
Thoreau, Henry David, 47, 78, 83, 100
Thornber, Karen, on emotion, 25–26
Thrailkill, Jane, 164

Tillett, Rebecca, 194, 260*n*50
Tone, 44, 67–68, 137, 140, 239*n*55; anxious, 169–70, 173–74, 204–5; apocalyptic, 201; of metaphor, 106–7; in *North Enough: AIDS and Other Clear-Cuts*, 47, 64–65; sexual, 62
Topophilia: A Study of Environmental Perception (Tuan), 22
Tourism, 64, 99; sandhill crane, 97, 113, 115–16
Toxic discourse, 21; Buell, L., on, 139
Toxic endangerment, 2
Toxicity, *see* Environmental toxicity; Pervasive toxicity
Toxification, 151; environmental imagination of, 2, 145, 147; representation of, 163–64; as signal, 145; of space, 131–32, 139; unbridled, 138; for Wallace, 137, 222; *see also* Contamination; Environmental toxicity; Pollution
Trauma, 46–47, 102, 104, 176, 191
Trauma studies, 23
Treichler, Paula, 41
Tropes: of bodies, 30, 48–49, 50, 140, 143–44, 199–200; in environmental writing, 221–22; form and, 128; master, 40, 44; medicalizing, 169, 197, 199–200, 204, 222; sickness, 3–4; *see also* Medicalization
Tuan, Yi-fu, 199; *Topophilia: A Study of Environmental Perception*, 22

Ugly Feelings (Ngai), 26
Urbanization, 5, 152
Urban planning, 59
Utopia, 172–73, 207, 221, 238*n*36; anxiety and, 173, 257*n*5; of everyday habit, 266*n*15; features of, 29; feminist, 168, 224; postapocalyptic, 215–16; realization of, 169–70; reviving, 238*n*36

Verghese, Abraham: adaptation of, 34; identity of, 37; medical research of, 31, 235n8; *My Own Country: A Doctor's Story*, 32–39, 67, 241n69; observations of, 35–38; PWAs and, 31–32; on rurality, 35–36; sentimentality of, 34, 45, 50

Wald, Priscilla, 7; *Contagious*, 20
Waldby, Catherine, 41, 191, 193
Walking, 11, 59; in *Infinite Jest*, 146, 150–52; *see also* Flâneur
Wallace, David Foster, 4, 7, 106, 222, 224, 228, 251n20, 254n94, 256n121; *The Broom of the System*, 133; on connectedness, 28163; "E Unibus Pluram," 124, 127, 158, 252n29; on fiction, 28, 126–27, 154, 158, 252n29; *Infinite Jest*, 11–12, 15, 17, 20, 26, 28–29, 108, 110, 116, 120–26, 128–31, 134, 138, 148–49; Kenyon College commencement speech, 164–65; new social arrangements, 133; *The Pale King*, 164, 255n101; style of, 153–54, 163–64, 251n24; suicide of, 154; toxification for, 137, 222; vision of, 29
Was (Ryman), 33
Waste, 48–51, 64–65, 130–34, 137, 252n35; consumption and, 130–31, 137, 138; displacement, 133, 138; dump, 122; industrial, 189; mining, 5, 180; prodigal, 130–31
Water, 92, 96–97, 114, 182, 186–89; depletion, 221; patterns of, 113; preservation, 115
West, the (American), 10, 56, 63, 182, 261n67, 262n71
Westling, Louise, 145
Weston, Kath, 33
White, Gilbert, 78
White Noise (DeLillo), 6
WHO, *see* World Health Organization
Wilde, Oscar, 236n22
Wilderness Act (1964), 8
Wilson, Edward O., 22, 233n63

Wojnarowicz, David, 4, 7, 20, 25–26, 45, 72; art of, 55–56, *57–58*; *Close to the Knives: A Memoir of Disintegration*, 27, 33, 34, 38–39, 43, 46, 55–70, 74–75, 222, 224, 240n58, 240n60, 240n62; death of, 55; Grover compared with, 63–65; losses of, 59; mobility of, 59–60; on nature, 60–62, 219; political commitments of, 72; prison metaphor of, 62; rhetoric of, 57, 59; tone of, 67–68; on violence, 60, 239n52; on woods, 60–61
Womack, Craig, 75
Woman on the Edge of Time (Piercy), 20, 168–75, 200, 221; agrarianism in, 207; biotechnology in, 29, 172–75, 185, 215; bodies in, 185–86; genetics in, 169, 173, 191, 194–95; healing in, 214; intimacy in, 215
Women, inequality and, 29
Wonder, 7, 22–23, 27, 76–79, 81–93, 120, 124, 219, 244n28, 244n30, 245n37; causality and, 80; defamiliarization and, 93, 115; in *The Echo Maker*, 28, 76, 79–81, 83, 88–95, 98, 100, 108, 114–15, 163, 223; in nature writing, 28, 78; pessimism and, 85, 116; structure of, 80; trajectories of, 28, 95–96
World Health Organization (WHO), on environmental illness, 1
World Wildlife Fund (WWF), 117–22, 124, 137, 153, 156, 165, 225
Writing for an Endangered World (L. Buell), 21
WWF, *see* World Wildlife Fund

Yoon, Carol Kaesuk, 78

Zoomorphism, technomorphism and, 111
Zunshine, Lisa, 102